OPTICAL–FIBER VELOCITY AND PRESSURE TRANSDUCERS

OPTICAL–FIBER VELOCITY AND PRESSURE TRANSDUCERS

V. G. Zhilin
*Institute for High Temperatures
Academy of Sciences of the USSR, Moscow*

Translated by

S. Chomet
King's College London

Edited by

A. J. Rogers
King's College London

○ HEMISPHERE PUBLISHING CORPORATION
A member of the Taylor & Francis Group
New York   Washington   Philadelphia   London

OPTICAL-FIBER VELOCITY AND PRESSURE TRANSDUCERS

Copyright © 1990 by Hemisphere Publishing Corporation. All rights reserved. Printed in the United States of America. Except as permitted under the United States Copyright Act of 1976, no part of this publication may be reproduced or distributed in any form or by any means, or stored in a data base or retrieval system, without the prior written permission of the publisher.

1 2 3 4 5 6 7 8 9 0   BRBR   8 9 8 7 6 5 4 3 2 1 0 9

This book was typeset by Karl Isitin and Company, London
Cover design by Sharon DePass.

Library of Congress Cataloging-in-Publication Data

Zhilin, V. G. (Viacheslav Gavrilovich)
  [Volokonno-opticheskie izmeritel 'nye preobrazovateli skoroski i davleniia. English]
  Optical-fiber velocity and pressure transducers / V. G. Zhilin : translated by S. Chomet : edited by A. J. Rogers.
    p.   cm.
  Translation of: Volokonno-opticheskie izmeritel 'nye preobrazovateli skorosti i davleniia.
  Includes bibliographical references.

  1. Fiber optics. 2. Displacement transducers. 3. Pressure transducers. I. Rogers, A. J. II. Title.
TA1800.Z4713   1990
621.36'92—dc20   89-36175
                                              CIP

ISBN 0-89116-308-5

CONTENTS

Preface   vii

Introduction   1

1 Optical-Fiber Displacement Transducers for Pressure and Velocity Sensors   11
    1 Elements of Lightguide Optics   17
    2 Fiber-Optical Displacement Transducers for a Reflecting Surface   15
    3 Optical-Fiber Displacement Transducers for Opaque Bodies   11
    4 Illuminating Lightguide Displacement Transducers   22

2 Optical-Fiber Pressure Transducers   29
    5 Optimization of Membrane Pressure Transducers   29
    6 Estimated Characteristics of Optical-Fiber Pressure Transducers with Glass Membranes   33
    7 Fabrication of Miniature Pressure Transducers and the Results of Tests Upon Them   35

3 Optical-Fiber Velocity Transducers   41
    8 Design of Sensitive Element   41
    9 Optical-Fiber Transducers of the Velocity of a Transparent Fluid   47
    10 Optical-Fiber Velocity Transducers for Opaque Fluids   64

11 Optical-Fiber Velocity Transducers in Flows with Velocity and Temperature Gradients ... 90

4 Dynamic Characteristics of Optical-Fiber Velocity and Pressure Transducers ... 97

12 Mathematical Description of the Oscillations of Sensitive Elements ... 97
13 Oscillations with a System of One Degree of Freedom ... 101
14 Dynamic Characteristics of Velocity Transducers for Transparent Liquids [19] ... 107
15 Dynamic Characteristics of Velocity Transducers for Opaque Liquids ... 122
16 Dynamic Characteristics of Optical-Fiber Pressure Transducers ... 130

5 Electronics for Optical-Fiber Transducers ... 139

17 Sources of Radiation for Optical-Fiber Transducers ... 139
18 Photodetecting Devices ... 142
19 The Microprocessor ... 150

Conclusion ... 157

References ... 159

Index ... 167

PREFACE

The flow velocity field is a major area of investigation because the rate of such important processes as heat and mass transfer is largely determined by the flow velocity field. Pressure fields are no less important, since they determine the dynamic effects of flows on bodies, and the amount of energy expended in fluid transport.

Most flows taking place in industrial systems and in nature (in the atmosphere and hydrosphere) are turbulent in character. Velocity and pressure exhibit stochastic fluctuations in both space and in time, which means that both time-averaged and instantaneous values of these quantities have to be measured.

The methods used to measure velocity and pressure must be capable of dealing with the complex character of these fields, and the primary sources of data (transducers) must have adequate sensitivity and the necessary resolution in space and time.

The time resolution is characterized by the maximum frequency component of the measured variable for which dynamic uncertainties are still small. Resolution in space can be characterized by the size of the volume within which the transducer takes the average of the measured variable. In the first approximation, spatial resolution is determined by the dimensions of the sensitive element of the transducer, especially in the direction of the gradient of the variable under investigation.

These conditions have governed the development of methods for velocity and pressure measurement during the last decade. This has led to the development of modern hot-wire and laser Doppler anemometers that can be used under relatively simple conditions (in gas and transparent-liquid flows in the presence of small temperature gradients). However, the more common situation is more complicated, that is, measurements have to be made in electrically conducting and opaque liquids. Moreover, there are still many unresolved methodological problems relating to the measurement of velocity in turbulent flows of liquid metals in magnetohydrodynamics and in the mechanics of electromotive flows. New methods of measurement that can match traditional techniques are therefore needed. This book shows that optical-fiber techniques offer a unique way of resolving these problems.

Fiber optics is a relatively new science and technology. The first monograph on this topic was published in the USSR in 1968 [64, 69]. Optical fibers are being used to detect small displacements of the light-reflecting membranes of sensitive pressure transducers.

The last decade has also seen the development of other types of optical-fiber transducers that transform displacement into electric signal and open new avenues for existing mechanical velocity transducers. These transducers are described on Chapter 1. They have applications beyond velocity measurement, and the results presented in this chapter may be of interest independently of the other chapters. This is why Chapter 1 also presents an account of optical-fiber transducers for the displacement of light-reflecting surfaces, other than those employed in membrane-type pressure transducers. The latter are described in Chapter 2 in which particular attention is devoted to the material used for the sensitive elements (membranes). It is shown that the use of thin membranes significantly simplifies fabrication and improves the characteristics of pressure transducers.

All the known designs of optical-fiber velocity transducers are presented in Chapter 3. It also provides a more detailed account of velocity transducers for opaque media, since they are promising and sometimes unique (for electromotive flows) devices that provide objective data (with known uncertainties) on turbulence in liquid-metal flows.

Chapters 2 and 3 describe the fabrication technology for optical-fiber velocity and pressure transducers. I hope that this will satisfy the needs of readers that are familiar with the transducers described in periodical literature, but who often ask for more detailed information.

Chapter 4 presents the methods used to calculate and measure the dynamic characteristics described in Chapters 2 and 3. These methods can be recommended not only for optical-fiber, but also any other mechanical elements in the form of cantilevers and membranes.

Chapter 5 describes the electronics used with velocity and pressure transducers. I hope that some of the recommendations relating to the choice of sources of light and very low frequency integrators may be found useful by experimenters.

Some explanation is needed of our use of the phrase "optical-fiber velocity and pressure transducers" in relation to mechanical transducers that have long been in use. We justify this use by the fact that fiber optics gives these devices qualitatively new properties. Other designations that may also be noted here are inductive, capacitive, piezoelectric, and other mechanical measuring transducers whose distinguishing feature is the conversion of the displacement (deformation) of a sensitive element into an electric signal.

Chapter 5 was written in collaboration with A. A. Oksman who developed and built the prototype electronics for transducers.

It is my pleasant duty to thank V. B. Ankudinov, Yu. P. Ivochkin, V. P. Ogorodnikov, and V. V. Osipov who took part in the development and testing of optical-fiber velocity and pressure transducers, and N. N. Tsypulev who helped with the preparation of the manuscript.

I am greatly indebted to V. G. Domrachev who refereed this monograph and to A. B. Shpanov who was the publishers' editor. Both of them provided valuable advice and suggestions that have resulted in significant improvements.

This book is the first systematic account of these new experimental techniques, and one naturally expects a response from readers on both fundamental and presentational questions. I should be grateful if readers would send their questions and suggestions to me via the publishers.

V. G. Zhilin

INTRODUCTION

Mechanical transducers are widely used to measure pressure, weight, acceleration, and other quantities that are readily transformed into forces that act on the elastic sensitive element (SE) of the measuring device in which they produce deformation (displacement). Tensometric, capacitive, inductive, piezoelectric, and other transducers are used to transform the (small) deformation of the SE into an electric signal [3, 34, 43, 61].

The mechanical fluid–velocity transducer is particularly attractive because velocity is simply transformed into the drag acting on the transducer head:

$$P = 0.5\, C_d(\text{Re}_d) \rho_f w^2 S$$

where $C_d(\text{Re}_d)$ is the drag coefficient which, for bodies of simple shape, is a known function of the Reynolds number $\text{Re}_d = wd/\nu_f$, ρ_f is the fluid density, ν_f is the kinematic viscosity, S is the effective cross sectional area of the body in the flow ($S = dl$ for a transversely placed cylinder and $S = \pi d^2/4$ for a sphere), and d and l are, respectively, the diameter and the length of the transducer head.

The only physical properties of the fluid that are present in the above formula are its density and viscosity. This means that the characteristics of a mechanical velocity transducer do not depend on the electrical conductivity, the thermal conductivity, and the specific heat of the medium, which substantially simplifies the process of measurement under non-isothermal conditions in the presence of electric and magnetic fields. It also means that there is an extensive range of fluids that are suitable for the investigation of flow and heat-transfer processes.

Mechanical velocity transducers have further advantages as compared, for example, with hot-wire anemometers. In particular, there is no limit on their sensitivity when time-independent velocities are measured, since the stiffness of the sensitive element can be made as low as desired. With suitably chosen materials, mechanical transducers can be used in aggressive or electrically conducting media (for example, electrolytes). Another significant advantage is that the working fluid need not be particularly pure. For example, it is well known that quantitative measurements are prevented by the presence on the surface of the hot wire in an anemometer, of a contaminant whose thickness amounts to a few percent of the wire diameter. This means that dust or other contaminating materials must be carefully removed from the flowing fluid [26]. A mechanical transducer is less sensitive to contaminants because the deposition of a contaminating material results simply in an increase in the size of the sensitive head. For a cylindrical head, we have, in the first approximation,

$$P \sim d^m$$

where d is the diameter of the head and m is a parameter that depends on the Reynolds number Re_d ($m \simeq 0.2$ for $\text{Re}_d < 1$ and $m \simeq 1$ for $\text{Re}_d > 1000$). Consequently, at worst, an increase in d by a few per cent produces a comparable percentage error in the measured drag.

There is a great variety of published designs of mechanical velocity transducers employing different geometries and different methods of transforming the deformation of the sensitive element into an electrical signal. Sensitive elements employed include elastic membranes and rods with attached receiving elements (heads) in the form of spheres, disks, slender bodies, and so on. The displacement of the sensitive element is transformed into an electrical signal by piezoelectric transducers [51], strain gauges [11, 15, 33, 49], differential capacitive or inductive transducers [12, 49], differential transducers of small displacements, filled with an electrolyte [27], and so on. Electrolytic transducers rely on the dependence of the electrical resistance between the electrodes and the separation between them. The gap between the electrodes is filled with an electrolyte with n-type conductivity [27]. Another example is the determination of the displacement of an opaque screen which overlaps a beam of light falling on a photoresistor [39].

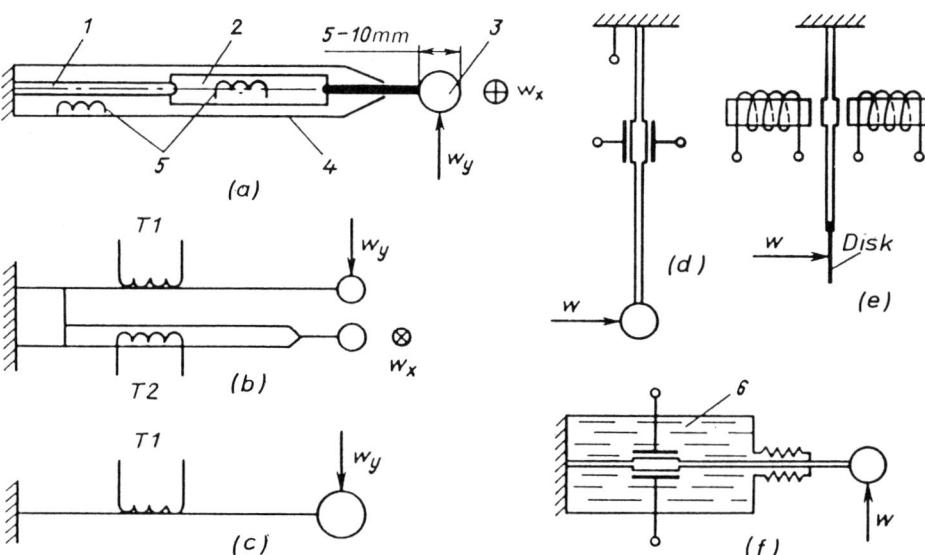

Fig.1 Mechanical velocity transducers: a, b, c — two— and one—component strain gauge type, c, d, e — capacitive, inductive, and electro—optical displacement transducers, 1, 2 — mutually perpendicular plates of sensitive element, 3 — ball head, 4 — shield, 5, T1, T2 — tensoresistors, 6 — cavity filled with electrolyte

Some designs of a mechanical velocity transducer are illustrated in Fig. 1. The two velocity components are measured by sensitive elements in the form of elastic plates arranged so that their stiffness in the direction of the measured velocity component is a minimum. Figure 2 [51] illustrates the principle of a transducer used to measure only the transverse component of turbulent velocity fluctuations. The transducer head is a symmetric miniature vane 1 attached to a conical cantilever 3. The vane is mounted at zero angle of attack to the mean velocity vector $<w_1>$ so that the lift P_2, which is proportional to the angle of attack α, is entirely due to the transverse component w'_2 of the velocity fluctuation. It is found that $w'_2/<w_1> \sim P_2$.

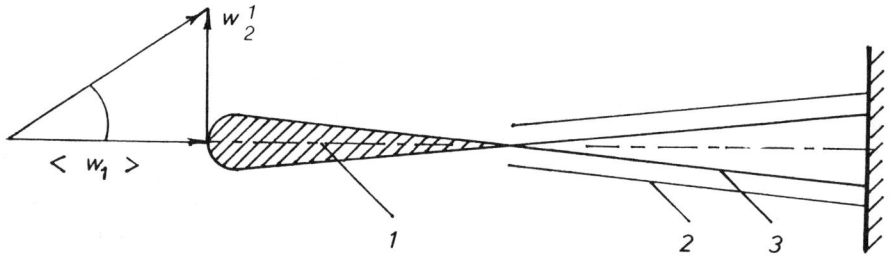

Fig.2 Measurement of the transverse component of velocity fluctuations

The lift P_2 is transformed into an electrical signal by a piezoelectric transducer mounted in the conical cantilever. The latter is surrounded by a jacket 2 which ensures that only the vane reacts to the turbulent fluctuations. Tests have shown that the spectrum of transverse velocity fluctuations recorded with this device is identical to that produced by the hot-wire anemometer in the frequency range 100 – 1000 Hz. The distortion of the transfer function at low frequencies is due to properties of the transducer whereas the distortion at high frequencies is due to the inertia of the mechanical part of the measuring system.

The above brief review of traditional methods of measuring fluid velocity with mechanical transducers is incomplete. We have failed to mention the different rotary transducers similar to the rotary anemometer, and many other designs more or less similar to those described above. Typically, they use a relatively large head (with linear dimensions up to a few millimeters), so that their spatial resolution is low and their dynamic characteristics unsatisfactory. The latter are not usually reported; at best, only the resonance frequency of the sensitive element is given and dynamic characteristics are estimated by comparing the time-dependent velocity measured by a mechanical transducer with measurements by other methods whose amplitude-frequency characteristics are arbitrarily assumed to be ideal [15, 51].

Analysis shows (see Section 8) that the miniaturization of mechanical velocity transducers can lead to a radical improvement in their metrological characteristics. However, this can only be done by overcoming the basic difficulties that are encountered when attempts are made to measure the very small forces that act on sensitive elements. When the dimensions of a cylindrical head in a mechanical transducer are comparable with the dimensions of the wire in a hot-wire anemometer ($l \simeq 1$ mm, $d \simeq 0.003$ mm), we find that a flow velocity of 1 m/s requires the transformation of a force of about 10^{-8} N into an electrical output signal. This seems unrealistic if the above designs are to be used to transform the force into a deformation, and the deformation into an electrical signal.

Optical methods of recording small deformations are therefore of considerable interest. Convective flow velocities were investigated in [67, 70]. The hot-wire anemometer or the Pitot tube could not be used because the velocity was too low and the flowing fluid was non-uniformly heated. The flow velocity was therefore measured by mechanical transducers in the form of thin quartz cantilever fibers whose deflection by the flow was determined with a traveling microscope. This type of transducer is very stable and its characteristics are relatively independent of the temperature of the fluid. However, the transducer measures only the time average of the velocity in a very special case (transparent fluid, relatively uniform flow, possibility of using a microscope). The transducer has been miniaturized by removing the deformation measuring system from the object under investigation.

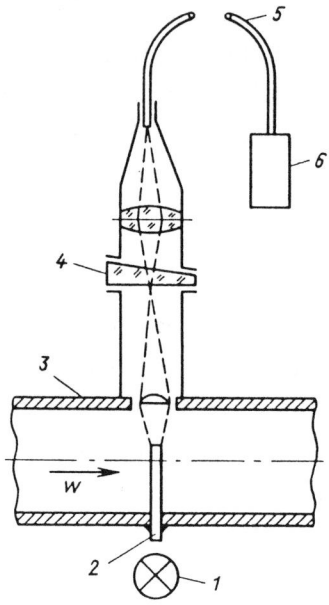

Fig. 3 Principle of optical—fiber anemometer

An important advance in the design of mechanical velocity transducers was made by A.N.Trokhanov [56] who proposed a device which he called the fiber anemometer. A simplified version of this transducer is shown in Fig.3. Light from the source *1* enters the quartz fiber *2*, attached to the wall of the tube *3* so that it is perpendicular to the direction of the fluid flow, which deflects the fiber. Light leaving the latter is displaced along the absorbing optical wedge *4* and is directed by the fiber bundle *5* on to the photomultiplier *6*. The photomultiplier output signal is thus a function of the deflection of the fiber, which in turn is uniquely related to the flow velocity. This device measures only one velocity component. Two perpendicular velocity components can be measured by using crossed optical fibers and a set of semitransparent mirrors. Trokhanov [56] discussed the possibility of a mobile probe incorporating optical fibers and light sources. However, difficulties were encountered with the adjustment and calibration of the probe in each new position.

Optical methods can thus be used to miniaturize sensitive elements and thus improve the metrological characteristics of mechanical velocity transducers. However, this is achieved by removing the system used to measure the deformation of the sensitive element from the transducer itself. This in turn means that it is difficult to measure time-dependent velocity fields, whilst the presence of temperature gradients can give rise to appreciable refraction effects because the light beam passes through a fluid with nonuniform optical properties.

Most of these difficulties can be obviated by using modern optical-fiber displacement transducers that are so small that they can be mounted together with sensitive elements [2 - 4, 13, 14, 37]. The velocity transducers designed in this way retain all the advantages of mechanical transducers but, in addition, have high spatial resolution and good dynamic properties. The development and properties of optical-fiber velocity transducers are discussed below. We also describe optical-fiber pressure transducer, first proposed about 20 years ago as an alternative to the then existing methods of measuring low, time-dependent pressures [5, 34, 43, 47, 61].

The pressure transducers used to measure pressure fluctuations in turbulent flows [52] have to satisfy particularly stringent requirements. Such measurements are possible only when the pressure transducer has high sensitivity, minimal size and high resonance frequency of the sensitive element.

These conditions are satisfied only in individual cases among widely used pressure transducers. For example, liquid-filled compensated manometers have sensitivities of up to 0.02 Pa in the pressure range 200-500 Pa, but can be used only for time-independent pressure measurement [43].

Membrane pressure gauges have reasonable sensitivity and frequency range [43, 47, 61], but their linear dimensions are typically of the order of 6–10 mm.

Piezoelectric transducers (linear dimensions of sensitive zone ~ 1 mm) have been successfully miniaturized, but their sensitivity is low and the typical range of measured pressures is 10^5 Pa [43, 61]. Moreover, piezoelectric transducers are unsuitable for measurements below 5 Hz. Other disadvantages include the loss of piezoelectric properties above a certain (relatively low) temperature. For example, in barium titanate this temperature is 124 °C. Other types of highly sensitive piezoceramics have a comparable upper working-temperature limit.

Capacitive transducers [34, 52] have very high sensitivity and good dynamic characteristics. The sensitive element is a thin (~ 2 μm) metal membrane, which forms a capacitor together with a flat electrode. Its capacitance depends on the electrode–membrane separation, so that the deflection of the membrane by the external pressure produces a change in the capacitance which, in turn, is recorded by suitable electronics. Measurements become simpler as the capacitance increases, since it is equal to the quotient of the membrane area and the membrane–electrode separation. This means that the linear dimensions of the system are relatively high (5 – 6 mm) and membrane–electrode separation is usually 15 – 20 μm. However, even with these parameters, the transducer capacitance is comparable with the capacitance of the leads, so that it is common to use frequency modulation of the signal, as well as other methods that complicate secondary equipment. It is also important to note that the calibration of these transducers is very dependent on the permittivity of the working medium, and it is difficult to protect the transducers from the effects of electromagnetic interference.

Searches for methods of transforming pressure into an electrical signal that would be free from the difficulties enumerated above have led to the use of the optical–fiber displacement transducer (Section 2) in

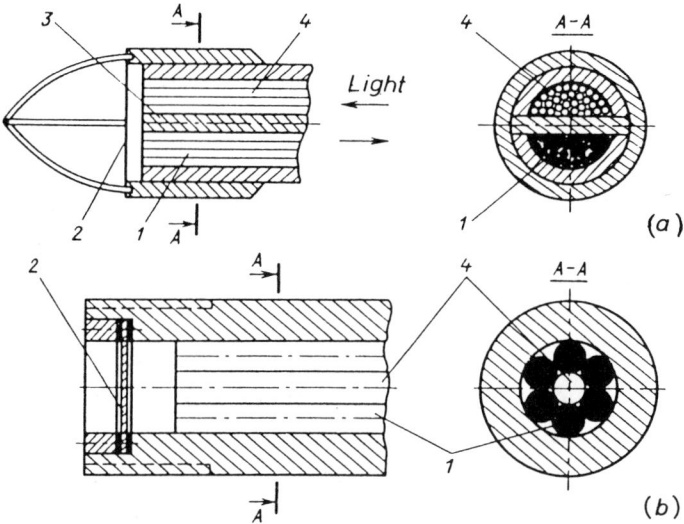

Fig.4 Optical–fiber pressure transducers

measurements of small displacements of sensitive elements (membranes). Even the earliest optical–fiber pressure transducers [64, 69] had very small linear dimensions (~ 1 mm) so that they could be introduced into the veins of animals as a means of measuring blood pressure. This is illustrated in Fig. 4a. The membranes *2* are 1 mm in diameter and 18 μm thick, and are made from a metallized polyester film. The illuminating lightguides *4* coupled to the source of light are separated from the receiving lightguides *1* by an opaque partition *3*. The ends of the two lightguide bundles are located at about 50 μm from the light–reflecting surface of the membrane. The light flux entering the receiving lightguides is then a function of the deflection of the membrane when a pressure acts upon it. Similar pressure transducers, but with metal membranes, are described in [4, 68]. The membranes used in [40] are made from polished aluminum and can be replaced (Fig.4b). Plastic lightguides, ~ 3 mm in diameter, are employed and are made in the form of a rosette in which the central lightguide illuminates the membrane and the peripheral lightguides collect the reflected light. Similar designs were proposed by the present author [68] and incorporated a stainless steel membrane.

None of the publications mentioned above is concerned with a detailed study of the metrological characteristics of optical-fiber pressure transducers. The only exception is [40] where an attempt is made to investigate the dynamic characteristics of an optical-fiber pressure transducer which is compared with a strain gauge in alternating pressure fields (up to 70 Hz). However, the results obtained do not provide a clear picture of the dynamic characteristics.

None of these publications explores fully the possibilities of optical-fiber pressure transducers. For example, there is great scope for increasing their sensitivity without loss of spatial resolution, or a deterioration in dynamical characteristics. The choice of the most suitable material for the sensitive element (membrane) has not been adequately discussed either.

CHAPTER
ONE

OPTICAL–FIBER DISPLACEMENT TRANSDUCERS FOR PRESSURE AND VELOCITY SENSORS

1. ELEMENTS OF LIGHTGUIDE OPTICS

Modern lightguides usually take the form of circular fibers made from optical glass, surrounded by glass cladding whose refractive index n_3 is smaller than the refractive index n_2 of the core material. Radiation propagates along the lightguide as a result of multiple total internal reflection of light rays at the boundary between the core and the cladding.

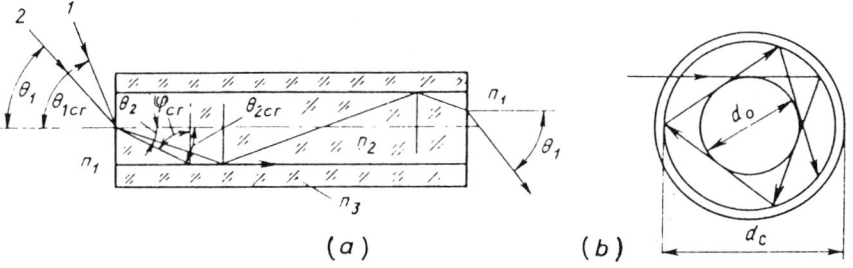

Fig.5 Ray paths in a circular lightguide: longitudinal (a) and transverse (b) sections

Consider rays propagating in a straight lightguide of constant circular cross section, with ends perpendicular to the optical axis (Fig. 5a). Ray *1* enters the lightguide at one end at an angle θ_{1cr} to the lightguide axis. After refraction, its angle to the axis becomes θ_{2cr}. It then strikes the core–cladding boundary at an angle ψ_{cr}, equal to the angle of total internal reflection. Any other ray *2* propagating at an angle $\theta_1 < \theta_{1cr}$ is totally internally reflected at the boundary, and eventually reaches another part of the surface at which it is reflected at the same angle, and so on. The ray thus propagates along the fiber, and eventually leaves it at the other end. The refraction of a ray of light at the separation boundary between two media with refractive indices n_1 and n_2 is described by Snell's law

$$n_1 \sin \theta_1 = n_2 \sin \theta_2$$

The condition for total internal reflection at the core–cladding boundary is

$$n_2 \sin \psi_{cr} = n_3$$

Since $\psi_{cr} = 90° - \theta_{2cr}$ it can be shown that the numerical aperture of the lightguide is

$$A = n_1 \sin \theta_{1cr} = (n_2^2 - n_3^2)^{1/2}$$

When a conical beam of rays is incident on one end of a perfectly transparent lightguide, and if the beam aperture angle is $\theta_1 < \theta_{1cr}$, the beam will be transmitted by lightguide with transmission coefficient $\tau = 1$. However, when $\theta_1 > \theta_{1cr}$, the light transmission coefficient under the same conditions is $\tau = \sin^2 \theta_{1cr}/\sin^2 \theta_1 < 1$. The transmission coefficient is $\tau = 1$ for any $\theta_1 \leqslant 90°$ if the refractive indices n_2 and n_3 are such that

$$A = (n_2^2 - n_3^2)^{1/2} \geqslant 1$$

All that we have said so far refers to meridional rays that cross the lightguide axis. For a given angle θ_1 at entry to the lightguide, oblique rays that do not cross the lightguide axis meet the core–cladding boundary at a greater angle than meridional rays. This means that, even for $\theta_1 > \theta_{1cr}$, some of the oblique rays will pass through the lightguide and will increase its light transmission. In contrast to meridional rays, oblique rays propagate along broken helical lines in the lightguide. The projection of an oblique ray on to the cross section of the lightguide is shown in Fig.5b. In a straight circular lightguide, the diameter d_0 of the cylinder touching a given ray remains constant as the ray propagates through the lightguide. The critical angle of incidence θ_{1cr}^0 for oblique rays is greater than the corresponding angle for meridional rays, and can be found from the formula for $n_1 = 1$:

$$\sin \theta_{1cr}^0 = A/(1 - d_0^2/d_c^2)^{1/2}$$

The angle θ_{1cr}^0 increases rapidly with increasing ratio d_0/d_c, reaching 90° for $n_1 = 1$, $n_2 = 1.7$, and $n_3 = 1.5$ when $d_0/d_c \sim 0.36$. The presence of oblique rays gives rise to an appreciable increase in the light transmission of the lightguide and in its effective aperture.

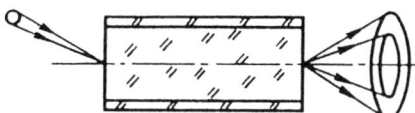

Fig.6 Symmetrization of rays in a lightguide

Properties such as the *symmetrization* of a beam of rays and *averaging* of illuminance over the exit aperture of the lightguide are important from the point of view of applications in lightguides and optical-fiber displacement transducers. Symmetrization is illustrated in Fig.6. A narrow conical beam is incident on one end of the lightguide and fills a zone defined by coaxial conical surfaces at the other end. Symmetrization of the rays inside the core and, hence, the displacement of the rays leads to the averaging of the distribution of light over the exit end, and to its uniform illumination. These properties of the lightguide are among the reasons for the linearity of the characteristics of optical-fiber transducers used to detect the displacement of an opaque body (see Section 3).

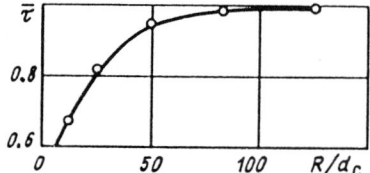

Fig. 7 Light transmission in a bent lightguide

Lightguides have a further very valuable property, namely, they have the ability to transmit light even when bent. It is clear from Fig.7 [29] that when the radius of curvature R of the lightguide does not exceed 6-7 lightguide diameters d_c, the relative light transmission $\bar{\tau}$ is about 60% as compared with the straight lightguide. So far, we have based our discussion on the laws of geometrical optics. This is valid when the diameter of the light-guiding core is several times greater than the wavelength of the transmitted radiation. However, as the core diameter approaches the wavelength, the lightguide begins to behave as a waveguide filled with a dielectric. Instead of the uniform illumination at the exit end, the intensity distribution takes the form of a symmetric pattern of bright and dark spots. Total internal reflection then occurs not at an arbitrary angle of incidence exceeding the critical values but only at certain discrete *characteristic* angles. Each such angle corresponds to a wave propagation mode and thus has its own intensity distribution at the exit end. As the lightguide diameter increases, the number of propagating modes increases, and the discrete spectrum of characteristic angles becomes denser, gradually becoming continuous. The intensity distribution at the exit end becomes uniform, in the limit.

2. FIBER–OPTICAL DISPLACEMENT TRANSDUCERS FOR A REFLECTING SURFACE

This type of optical–fiber displacement transducer has been known for at least twenty years [38, 40, 64, 65, 66, 68, 69]. Its principle is simple. The ends of the illuminating (*2*) and receiving (*4*) lightguides, which are coupled to the source of light *1* and photodetector *5*, respectively, are arranged so that they are parallel to the reflecting surface *3* (Fig. 8). The photodetector signal depends on the separation h between the reflecting surface and the ends of the lightguides. It is clear that when $h = 0$ or $h \to \infty$, light will not reach the receiving lightguide. Consequently, the dependence of the photodetector output u on the distance h will be a curve with a maximum.

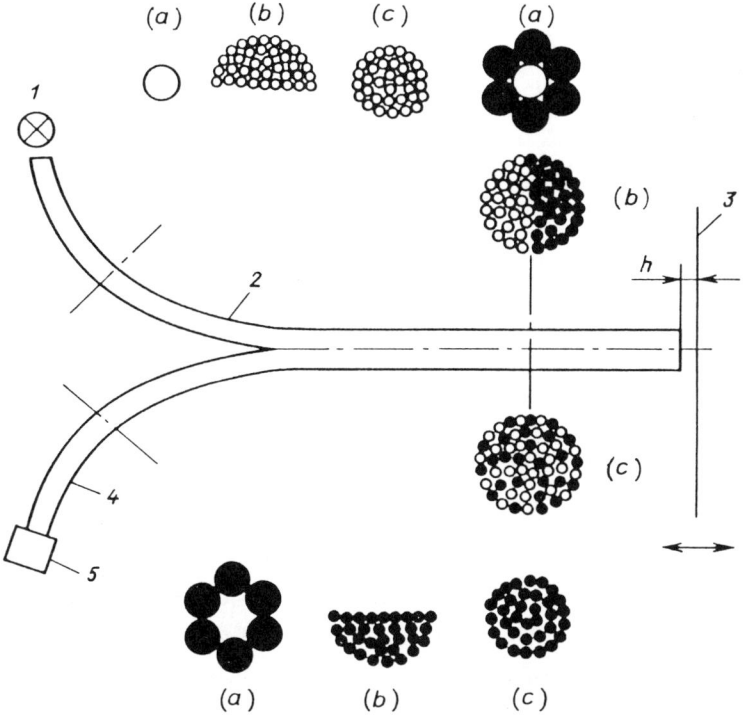

Fig.8 Principle of optical–fiber transducer for the displacement of a reflecting surface

There are different ways of combining illuminating and receiving lightguides (*a, b, c* in Fig.8) [38, 65, 69]. Evidently, the simplest and most effective is method *c* [38, 65] which employs a standard fiber bundle with an irregular arrangement of individual fibers. The ends of this bundle (diameter 1 mm) are glued together and polished. However, individual fibers can be separated with a sharp knife, so that some of them can be used for illumination and the rest for reception. A typical characteristic of this type of displacement transducer is shown in Fig. 9, in which

$$\bar{u} = (u - u_0)/u_0$$

where u_0 is the signal for $h = 0$, due to the photodetector dark current.

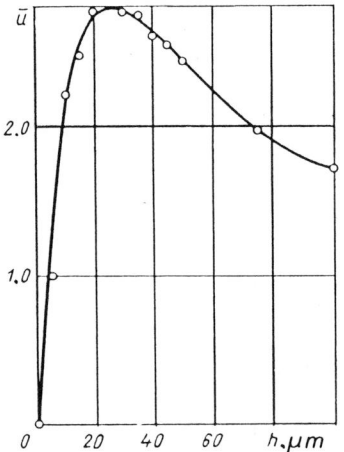

Fig.9 Calibration curve of an optical–fiber transducer for the displacement of a reflecting surface

The dependence of the signal on h has a maximum at $h \simeq d_r$ where d_r is the diameter of the fibers in the receiving bundle. This characteristic was recorded with a split bundle consisting of a bundle of fibers 0.025 mm in diameter. The position of the maximum at $h \simeq d_r$ is also confirmed for other fiber diameters in the bundle and for other fiber configurations (Fig.8a). The maximum displacement sensitivity (about 0.1 μm) is achieved on the ascending part of the characteristic shown in Fig.9. It is this part that is used in miniaturized optical-fiber high-sensitivity pressure transducers (see Sections 5 and 6). It is, of course, possible to use the descending part of the characteristic when displacements covering a wider range have to be determined. Of course, the displacement sensitivity is then lower.

3. OPTICAL-FIBER DISPLACEMENT TRANSDUCERS FOR OPAQUE BODIES

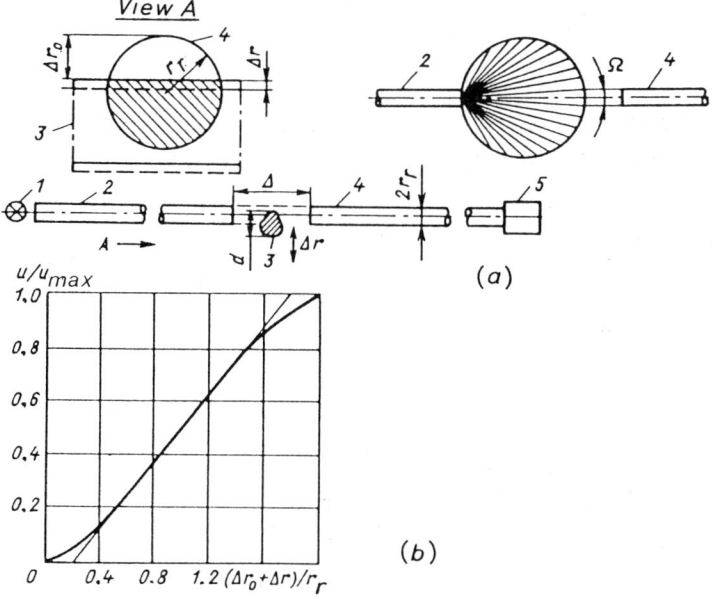

Fig.10 One-component, optical-fiber transducer for the displacement of an opaque body (a) and its characteristic (b)

Let us now consider two coaxial lightguides *2* and *4* (Fig.10a) with the opaque body *3* placed in the gap Δ between the parallel lightguide ends [37]. When the illuminating lightguide *2* is connected to a light source *1*, and the receiving lightguide *4* (radius r_r) is connected to photodetector *5*, the signal

u from the latter will depend on the displacement Δr of body 3 in the direction perpendicular to the optical axis of this system. This relationship is single-valued when $d > 2r_r$. Let us suppose that light is distributed uniformly over the cross section of the beam reaching the receiving lightguide. The light flux intercepted by the photodetector is then proportional to the unshaded area of the receiving lightguide cross section (Fig.10a). If at the same time, the photodetector output is proportional to the light flux, the dimensionless photodetector output is given by

$$\overline{u} = u/u_{max} = [\arccos(1 - \overline{\Delta r}) - (1 - \overline{\Delta r})(2\overline{\Delta r} - \overline{\Delta r}^2)^{1/2}]/\pi \qquad (1)$$

where u_{max} is the maximum value of u and $\overline{\Delta r} = (\Delta r_0 + \Delta r)/r_r$.

The function $\overline{u}(\overline{\Delta r})$ calculated from (1) is shown in Fig.10b and is almost linear in the range $0.5 < \overline{\Delta r} < 1.3$. Since Δr_0, which depends on the adjustment of the system, is usually not known accurately, it is more convenient to use the ratio $\overline{u} = (u - u_0)/u_0$ as a function of $\overline{\Delta r} = \Delta r/r_r$, taking $\overline{\Delta r_0} = \Delta r_0/r_r$ as the parameter. Subject to the above assumptions about the uniformity of the beam of light and the linearity of the photodetector, it then follows from (1) that

$$1 + \overline{u} = \frac{\arccos(1 - \overline{\Delta r_0} - \overline{\Delta r}) - (1 - \overline{\Delta r_0} - \overline{\Delta r})D}{\arccos(1 - \overline{\Delta r_0}) - (1 - \overline{\Delta r_0})(2\overline{\Delta r_0} - \overline{\Delta r_0}^2)^{1/2}} \qquad (2)$$

where $D = [2(\overline{\Delta r_0} + \overline{\Delta r}) - (\overline{\Delta r_0} + \overline{\Delta r})]^{1/2}$.

Figure 11 (points) shows the results of a calculation based on (2) for $\overline{\Delta r_0}$ between 0.6 and 1.4. It is clear from the Figure that, with the necessary initial adjustment ($0.6 < \overline{\Delta r_0} < 1.4$), this displacement transducer will introduce a slight nonlinearity into the transfer function of a device in which it is employed.

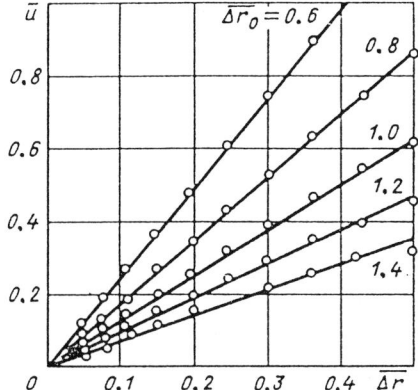

Fig.11 Output signal of a one−component displacement transducer for different initial displacements $\overline{\Delta r_0}$

As noted above, this conclusion is valid only under certain particular conditions. The photodetector is a linear device when it takes the form of a photomultiplier and a semiconductor photocell [24. 54]. The uniformity of the measuring beam depends on the angular properties of the illuminating lightguide and on the separation Δ between the ends of the illuminating and receiving lightguides. However, the most important quantity is the solid angle $\Omega \simeq \pi r_r^2/\Delta^2$, since it is clear that the smaller the solid angle Ω the more uniform the measuring beam.

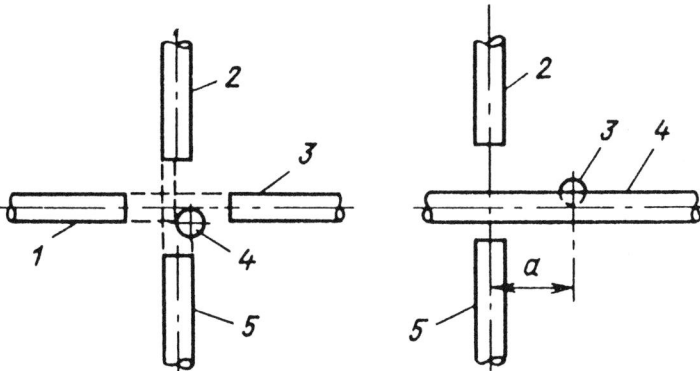

Fig.12 Two−component transducer for the displacement of an opaque body

The two components of the displacement of an opaque body (Fig.12) can be measured relatively simply by using two crossed transducers of the above type [13, 14]. The illuminating lightguides *1* and *5* are coupled to a source of light and the receiving lightguides (*2* and *3*) are coupled to two photodetectors. The system is adjusted so that the opaque body *4* partially interrupts the two measuring beams in its zero position. Two conditions must then be satisfied if the two displacement components are to be measured independently. First, the signal from each of the transducers must not depend on the displacement of the body along its optical axis. Second, the measuring channels must not illuminate one another during the displacement of the body. Whether or not the first condition is satisfied when the second is satisfied depends only on the uniformity of the measuring beam, and can be checked experimentally for each channel. This can be done by keeping constant the magnitude and direction of the displacement and rotating the transducer, in the plane of the drawing in Fig.12, around the axis cutting the optical axis of the transducer. When the transducer signal is independent of the displacement of the opaque body along the optical axis, the dimensionless signal $\bar{u}$ as a function of the angle of rotation α, measured from the direction of the displacement Δ*r*, is as follows:

$$\bar{u} = [u(\alpha) - u_0]/(u_{max} - u_0) = \sin \alpha \tag{3}$$

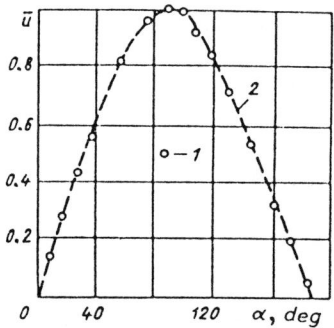

Fig.13 Experimental verification of formula (3)

This formula has frequently been verified experimentally. The results of one such verification are shown in Fig.13 [37] (*1* - experiment, *2* - calculated from the above formula). The above procedure can also be used to determine the angle between the optical axes of crossed displacement transducers. The required angle is determined from the phase difference between two sinusoids (see Fig.42).

As far as the mutual illumination of the two channels is concerned, this can be avoided by placing the optical axes of the transducers in different planes, perpendicular to the axis of the opaque body (see Fig.12). The distance a that ensures the independent operation of the two measuring channels is determined by the diameters of the lightguides, the separation between the ends of the illuminating and receiving lightguides, and the properties of the surrounding surfaces. It is chosen experimentally in each particular case. It is usually sufficient to ensure that $a > \Delta$.

The displacement sensitivity of the above transducers depends on many factors, including the source intensity, the lightguide diameter, the solid angle Ω, the photodetector sensitivity, and the properties of the electronic circuits; it cannot be determined from general considerations. We should therefore confine our attention to a few relevant points.

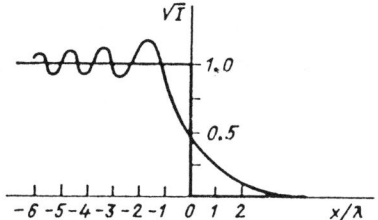

Fig.14 Diffraction at the edge of a screen

Two-component optical-fiber transducers for the displacement of an opaque body are used in velocity transducers (see Sections 9 and 10). The diameter of the receiving lightguides is between 15 μm and 150 μm. The displacement sensitivity is then calculated from the output velocity-signal characteristic and is found to be up to 0.1 μm. The calibration curves of the velocity transducers are found to be monotonic and reproducible. We must therefore consider how the transducer characteristic is affected by diffraction phenomena. Figure 14 shows the intensity I behind an opaque perfectly conducting screen, illuminated normally by a plane E-polarized monochromatic wave [7]. (In the E-polarized wave, the electric field vector is parallel to the edge of the screen.) It is clear from the figure that the intensity falls monotonically in the region of the geometric shadow, but elsewhere oscillates with the period close to the wavelength of the incident light. This can give rise to a nonmonotonic calibration curve when coherent monochromatic sources (lasers) are employed. These undesirable features of calibration curves have not been observed because unpolarized sources with a continuous spectrum (hot-filament lamps or light diodes) were employed. The displacement sensitivity can then be much less than the characteristic wavelength of the radiation.

The tendency toward miniaturization of displacement transducers has led to a reduction in the diameters of the illuminating and receiving lightguides in which the reduction in sensitivity is compensated by the higher source intensity. Both calculations and experiments show that reliable detection of the light beam transmitted by the transducer is possible even for lightguide diameters of about 0.001 mm. However, it is important to remember that waveguide effects occur when the lightguide diameter is comparable with the wavelength, and the measuring beam has a complicated nonuniform mode structure (see Section 1). The transducer characteristic then becomes nonmonotonic and time-dependent.

4. ILLUMINATING LIGHTGUIDE DISPLACEMENT TRANSDUCERS

In some designs of velocity transducers, it is convenient to use a lightguide as the elastic sensitive element (see Section 10) and to measure one of the components of its displacement with the transducer shown in Fig.15a [3]. It is clear that the signal from photodetector 1 depends on the separation Δ

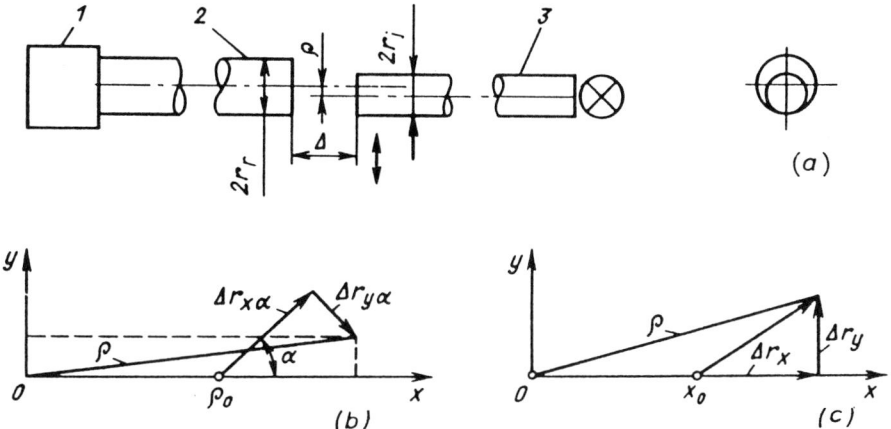

Fig.15 One—component transducer for the displacement of the illuminating lightguide

between the ends of the lightguides and the distance ρ between the axis of the illuminating 3 and receiving 2 lightguides. If we keep one of these quantities (Δ or ρ) constant, we can determine the other. Let us consider the ρ-transducer in greater detail, since it has been used in optical-fiber velocity transducers. Consider Fig. 15b. Suppose that the center of the cross section of the receiving lightguide is the origin of coordinates, and the center of the illuminating lightguide is initially at $x = \rho_0$, $y = 0$. We also know the direction of the displacement $\Delta r_{x\alpha}$ to be measured, and there is also a certain orthogonal displacement $\Delta r_{y\alpha}$ whose influence on the measurement of $\Delta r_{x\alpha}$ must be minimized. We have to find the optimum angle α between the displacement $\Delta r_{x\alpha}$ and the direction $O\rho_0$ in which sensitivity to the displacement $\Delta r_{x\alpha}$ is a maximum whereas sensitivity to $\Delta r_{y\alpha}$ is a minimum. This type of problem is encountered quite frequently in measurement practice. For example, when the velocity in a turbulent flow is measured with a hot-wire anemometer, both velocity components cannot be measured with the single wire. One therefore measures the component in the direction of the mean velocity. Other things being equal, the precision of this measurement depends only on the velocity component perpendicular to the direction of the mean velocity.

The transducer signal depends on ρ, so that we must find the conditions under which this quantity has maximum sensitivity to $\Delta r_{z\alpha}$ and minimum sensitivity to $\Delta r_{y\alpha}$. We have

$$\rho^2 = \left(\rho_0 + \Delta r_{z\alpha} \cos\alpha + \Delta r_{y\alpha} \sin\alpha\right)^2 + \left(\Delta r_{z\alpha} \sin\alpha - \Delta r_{y\alpha} \cos\alpha\right)^2 \quad (4)$$

We define the sensitivity to displacement $\Delta r_{z\alpha}$ by

$$S_{z\alpha} = \partial(\rho^2)/\partial(\Delta r_{z\alpha})$$

and, correspondingly,

$$S_{y\alpha} = \partial(\rho^2)/\partial(\Delta r_{y\alpha}).$$

Using (4), we then have

$$S_{z\alpha} = 2\rho_0 \cos\alpha + 2\Delta r_{z\alpha}$$

$$S_{y\alpha} = 2\rho_0 \sin\alpha + 2\Delta r_{y\alpha}$$

Hence, it follows that, when $\alpha = 0$, we have the maximum value $S_{z\alpha} = 2\rho_0 + 2\Delta r_{z\alpha}$ and the minimum value $S_{y\alpha} = 2\Delta r_{y\alpha}$.

Figure 15c shows the optimum orientation of the Δr_z transducer. Let us now determine the uncertainty in the measured Δr_z as a function of Δr_y for the optimum orientation. When $\alpha = 0$, we have

$$\rho^2 = (x_0 + \Delta r_z)^2 + \Delta r_y^2 \quad (5)$$

Substituting $\rho/x_0 = \bar{\rho}$ and $\Delta r_y/\Delta r_z = k$, we find from (5) that the relative uncertainty $\delta(\Delta r_z)$ due to the displacement component Δr_y, which we cannot control, is given by

$$\delta(\Delta r_z) = 1 - \frac{P-1}{Q}$$

where $P = \left[\left(1+k^2\right)\bar{\rho}^{-2} - k^2\right]^{1/2}$ and $Q = \left(1+k^2\right)\left(\bar{\rho}-1\right)$

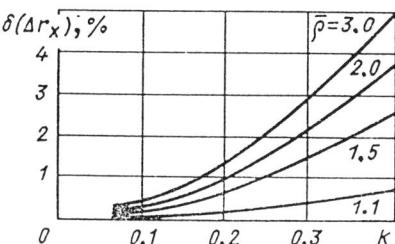

Fig.16 Uncertainty in Δr_z due to Δr_y

Figure 16 shows graphs of this expression. It is clear that this uncertainty is relatively small and decreases with increasing x_0 (decreases with $\bar{\rho}$).

If we add a second receiving lightguide (Figure 17a) to the system shown in Figure 15a, we can measure the two components of the displacement of the illuminating waveguide [14]. The center of the ends of the illuminating lightguide (Figure 17b) in its initial position is at x_0, y_0, and the axes of the receiving lightguides lie at $(0,0)$ and $(2r_r, 0)$. Suppose that the illuminating lightguide is displaced by the amount Δr to a point x, y. This displacement has two orthogonal components, namely, $\Delta r_{z\alpha}$ in the direction at an angle α to the x axis and $\Delta r_{y\alpha}$. From now on, we shall use only dimensionless quantities (with $2r_r$ as the unit of length). It is clear from the figure that the following expressions are valid:

$$\rho_1^2 = \left(x_0 + \Delta r_{z\alpha}\cos\alpha + \Delta r_{y\alpha}\sin\alpha\right)^2 +$$

$$+ \left(y_0 - \Delta r_{z\alpha} \sin\alpha + \Delta r_{y\alpha} \cos\alpha \right)^2 \tag{6}$$

$$\rho_2^2 = \left(1 - x_0 - \Delta r_{z\alpha} \cos\alpha - \Delta r_{y\alpha} \sin\alpha \right)^2 +$$

$$+ \left(y_0 - \Delta r_{z\alpha} \sin\alpha + \Delta r_{y\alpha} \cos\alpha \right)^2 \tag{7}$$

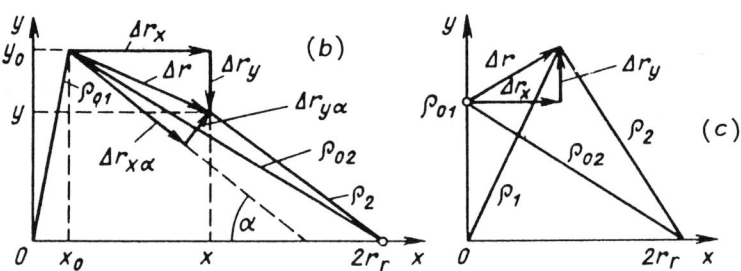

Fig.17 Two−component transducer for the displacement of the illuminating lightguide: *1, 6* − photodetectors, *2, 5* − receiving lightguides, *3* − illuminating lightguide, *4* − source of light

Let us suppose that the dimensionless photodetectors outputs are $\bar{u}_1$ and $\bar{u}_2$, and that they are related to ρ_1 and ρ_2 by certain functions:

$$\rho_1^2 = f_1(\bar{u}_1)$$

$$\rho_2^2 = f_2(\bar{u}_2)$$

$$\rho_1^2 - \rho_2^2 = f_1 - f_2 = \Delta f$$

It then follows from (6) and (7) that the simplest way of determining $\Delta r_{z\alpha}$ is to measure Δf for $\alpha = 0$. We then have $\Delta f = 2x_0 - 1 + 2\Delta r_{z\alpha}$. It follows that, as far as the determination of $\Delta r_{z\alpha}$ independently of $\Delta r_{y\alpha}$ is concerned, it is best to take $\alpha = 0$.

As far as $\Delta r_{y\alpha}$ is concerned, the displacement $\Delta r_{y\alpha}$ cannot be determined in terms of a similarly simple combination of functions f_i for any value of α, so that it must be found, for example, from (6).

Let us now analyze the effect of α on the sensitivity of the transducer, and find the conditions that ensure maximum sensitivity to both $\Delta r_{z\alpha}$ and $\Delta r_{y\alpha}$ [20]. We define the $\Delta r_{z\alpha}$ sensitivity by

$$S_{z\alpha}^{\Delta f} = \partial(\Delta f)/\partial(\Delta r_{z\alpha})$$

and the sensitivity to $\Delta r_{y\alpha}$ by

$$S_{y\alpha}^{f_1} = \partial f_1/\partial(\Delta r_{y\alpha})$$

It then follows from (6) and (7) that

$$S_{z\alpha}^{\Delta f} = 2\cos\alpha$$

$$S_{y\alpha}^{f_1} = 2\rho_{01}\cos[\alpha - \arcsin(x_0/\rho_{01})] + 2\Delta r_{y\alpha}$$

i.e., the $\Delta r_{z\alpha}$ sensitivity is a maximum for $\alpha = 0$. The quantity $S_{y\alpha}^{f_1}$ is also a maximum under these conditions if, at the same time, $x_0 = 0$. Moreover, $S_{y\alpha}^{f_1}$ increases with increasing $\rho_{01} = y_0$. The maximum sensitivity of the two-component transducer to both displacement components is achieved for $\alpha = 0$ and $x_0 = 0$, and this conclusion is independent of the form of functions f_i. The optical orientation of the transducer relative to the measured displacement components is shown in Fig.17c.

The increase in $S_y^{f_1}$ with increasing ρ_{01} does not mean, however, that we must always try to increase ρ_{01}. This depends on the form of the functions f_i. The dependence of $\bar{u}_i = u_i/u_{i\,max}$ on ρ_i^2, where $u_{i\,max}$ corresponds to $\rho_i = 0$, is nearly exponential:

$$\bar{u}_i = \exp(-\alpha \rho_i^2)$$

where a decreases with increasing Δ.

Let us suppose that the calibration curves $\bar{u}_i = \phi_i(\rho_i^2)$ have been linearized, i.e.,

$$\bar{u}_i = 1 - \rho_i^2/\rho_{max}^2$$

$$\Delta f = \rho_1^2 - \rho_1^2 = -\rho_{max}^2 (\bar{u}_1 - \bar{u}_2)$$

$$f_1 = -\rho_{max}^2 (1 - \bar{u}_1)$$

$$S_z^{\Delta \bar{u}} = \partial(\bar{u}_1 - \bar{u}_2)/\partial(\Delta r_z) = -S_z^{\Delta f}/\rho_{max}^2 = -2/\rho_{max}^2$$

$$S_y^{\bar{u}_1} = \partial \bar{u}_1/\partial(\Delta r_y) = S_y^{f_1}/\rho_{max}^2 = -2\rho_{01}/\rho_{max}^2 - 2\Delta r_y/\rho_{max}^2$$

where ρ_{max} is the value of ρ_i for $\bar{u}_i = 0$.

The sensitivity to Δr_z is then found to increase with decreasing ρ_{max}. As far as $S_y^{\bar{u}_1}$ is concerned, there are optimum values of ρ_{01} and ρ_{max}. Since $\rho_{max,\,min}^2 = 1 + \rho_{01}^2$, we find that

$$S_y^{\bar{u}_1} = 2\rho_{01}/(1 + \rho_{01}^2)$$

is a maximum for $\rho_{01} = 1$ and $\rho_{max} = \sqrt{2}$. Moreover, $0 \leq \Delta r_z \leq 1$, and this defines the range of displacements in which this transducer is used.

CHAPTER
TWO
OPTICAL–FIBER PRESSURE TRANSDUCERS

5. OPTIMIZATION OF MEMBRANE PRESSURE TRANSDUCERS

In general, the displacement z_0 of the center of a peripherally-clamped circular membrane and the applied pressure difference p across it are related by [47]

$$\frac{pr_0^4}{Eh^4} = \frac{A_1}{1-\mu^2}\frac{z_0}{h} + A_2\left(\frac{z_0}{h}\right)^3 \tag{8}$$

where r_0 and h are, respectively, the radius and thickness of the membrane, E and μ are Young's modulus and Poisson's ratio of its material, A_1 and A_2 are constants independent of the parameter m in the equation $\theta = B\,(\bar{r}^{-m} - \bar{r})$ of the deformed surface of the membrane, θ is the angle between the normal to the meridional cross section of the membrane and its axis of symmetry, and $\bar{r} = r/r_0$ is the dimensionless radius. When $z_0/h \ll 1$ (small deflections), $m \simeq 3$, and $A_1 = 16/3$, equation (8) assumes the simplified form

$$z_0 = \frac{3}{16} p \frac{1-\mu^2}{Eh^3} r_0^4 \qquad (9)$$

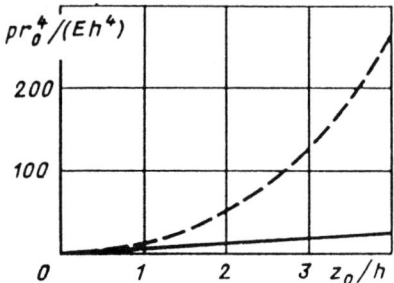

Fig.18 Comparison between calculations based on (8) and (9)

Equation (8) must be used whenever it is necessary to perform an exact calculation of the sensitivity of the pressure transducer from known values of p, r_0, h, E, and μ, with the parameters m, A_1 and A_2 determined by the method described and verified experimentally in [47]. The difference between (8) and (9) can be judged by inspection of Fig. 18 (dashed and solid curves, respectively) in which these two equations are compared for $z_0/h < 4$. The reason for the significant increase in the stiffness of the membranes for large relative deflections is the increasing role of tensile strain as the deflection increases.

We shall confine our attention to a qualitative analysis of the membrane transducer on the basis of (9), since it is only in this case that the necessary relationships can be obtained in a closed form.

As already noted, the ideal pressure transducer must have high sensitivity, minimum size, and maximum possible membrane resonance frequency (see Section 16):

$$f_r = \frac{1.625 h}{r_0^2} \sqrt{\frac{E}{12\rho(1-\mu^2)}} \qquad (10)$$

where ρ is the density of the membrane material.

It then follows from (9) that

$$r_0 \sim \left(\frac{z_0}{p} \frac{E}{1-\mu^2} h^3 \right)^{1/4}$$

Consequently, to minimize r_0 and thus increase the spatial resolution, we must try to reduce z_0/p, h, and $E/(1 - \mu^2)$. The value of z_0/p is wholly determined by the sensitivity of the displacement transducer. The more sensitive the transducer, the smaller the value of z_0/p that is necessary from the point of view of precision and the greater the necessary miniaturization of the pressure transducer. It is therefore desirable to use the ascending branch of the characteristic of the optical–fiber displacement transducer (see Fig.9).

Table 1

Material	$E \times 10^{11}$, N/m²	μ	$\rho \times 10^3$, kg/m³	$[E/(1-\mu^2)]^{1/4}$, N$^{1/4}$ m$^{-1/2}$	$\bar{f}_p$
Glass	0.60	0.25	2.6	503	1.00
Aluminum	0.70	0.34	2.7	530	0.96
Steel	2.20	0.28	7.8	706	0.57
Silver	0.77	0.37	10.5	546	0.49
Gold	0.80	0.42	19.3	558	0.36
Platinum	1.70	0.39	21.4	669	0.31

The value of r_0 is also greatly influenced by the thickness h of the membrane. Consequently, the membrane material must be technologically acceptable and must be such that thin membranes can be fabricated in a relatively simple way. The quantity $E/(1 - \mu^2)$, which also affects r_0, can be reduced to a minimum by a suitable choice of the material. Table 1 lists the properties of a number of materials that can be used to fabricate thin membranes. It is clear from the Table that glass and aluminum have the most suitable combinations of E and μ.

Since the necesssary value of z_0/p is wholly determined by the sensitivity of the displacement transducer, and can be regarded as given within the framework of our present analysis, we have

$$r_0 \sim E^{1/4} (1 - \mu^2)^{-1/4} h^{3/4}$$

It then follows from (10) that

$$f_r \sim \rho^{-1/2} h^{-1/2}$$

Consequently, for the minimum value of h (based on technological considerations), the resonance frequency of the membrane is higher, and its material better, the lower the density ρ. It is clear from the Table that glass and aluminum are the most suitable materials from this point of view. Other things being equal, the values of f_r for the glass membrane are greater by a factor of almost two than for the steel membrane.

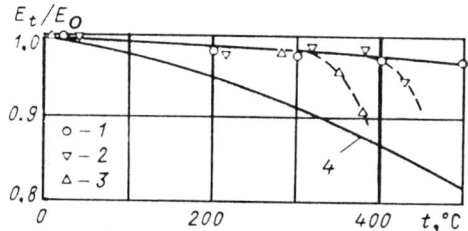

Fig.19 Young modulus of glass and steel as a function of temperatures: 1 — soda—free glass, 2 — potassium—sodium glass, 3 — sillicate—lead glass, 4 — steel

Glass also has certain other advantages as compared with metals, namely, it is technologically more convenient and its mechanical properties are not very temperature dependent. Figure 19 shows the Young's modulus of glass [23] and of steel as functions of temperature. As can be seen, Young's modulus of different glasses is practically independent of temperature in the range 300–600 °C (i.e., between room temperature and the softening point). The Young's modulus of steel, on the other hand, falls by almost 20% in the same range. The mechanical properties of aluminum are even more sensitive to temperature.

Because of the strong temperature dependence of their mechanical properties, and also for technological reasons, certain polymeric materials that might have competed with glass are not listed in Table 1.

Glass is a dielectric, so that the use of glass membranes in, for example, capacitive transducers, requires the deposition of conducting (metal) coatings. This can give rise to an increase in the effective density of the membrane and to a deterioration in its dynamic characteristic. The great advantage of optical-fiber pressure transducers is that they do not depend on the electrical conductivity of membranes. When the medium on both sides of the transparent membrane is air (or some other gas), the minimum value of the reflection coefficient at normal incidence [7] is

$$R_{min} = 2\left[(n-1)/(n+2)\right]^2$$

where n is the refractive index of the membrane material (glass). For $n = 1.5$, we then have $R_{min} \simeq 0.04$. Experiment shows that the light intensity reflected from the uncoated glass membrane is sufficient for the reliable operation of the optical-fiber displacement transducer if we use a bundle consisting of a large number of irregularly packed lightguide fibers (see Fig.8c).

6. ESTIMATED CHARACTERISTICS OF OPTICAL-FIBER PRESSURE TRANSDUCERS WITH GLASS MEMBRANES

When we estimate the basic transducer characteristics we shall assume that the optical-fiber system used to measure the deflection of the membrane consists of optical fibers 25 μm in diameter. To ensure maximum sensitivity, a single-valued calibration curve, and maximum pressure range, the end of the optical fiber bundle must be placed near the maximum of the characteristic (Fig.9), i.e., at a distance from the membrane equal to the fiber diameter. We shall take this distance to be 25 μm, although it is possible to use finer fibers, thus reducing the separation between the end of the cable and the membrane and increasing the sensitivity of the transducer. However, this is accompanied by a reduction in the measured-pressure range, and gives rise to adjustment difficulties and to a higher probability that the membrane will be damaged when the fiber cable is mounted.

Having set the distance between the end of the cable and the membrane, we must determine the maximum possible membrane deflections $z_{0\,max}$ and then use (9) and (10) to determine the relationship between the maximum measured pressure p_{max}, membrane thickness h, the membrane radius

$$r_0 = \left[\frac{16}{3} \frac{E z_{0\,max} h^3}{(1-\mu^2) p_{max}} \right]^{1/4}$$

and the resonance frequency

$$f_r = 0.20 \left[\frac{p_{max}}{\rho h z_{0\,max}} \right]^{1/2}$$

Figure 20 shows the calculated ρ_0, f_r, and σ_{max} for three values of h and values of p_{max} reaching 10^4 Pa. It is clear from the figure that, for example, at $p_{max} = 1000$ Pa ($\simeq 100$ mm of water column) and $h = 1$ μm, the required membrane diameter is $2r_0 \simeq 0.6$ mm, and the minimum resonance frequency reaches 30 kHz. The value of σ_{max} is then found to lie between reasonable limits ($\simeq 70$ N/mm^2).

It is important to remember that the results of this calculation are more likely to characterize the limiting attainable characteristics of optical-fiber pressure transducers because, in practice, it is difficult to achieve a constant membrane thickness over its entire area, and to ensure

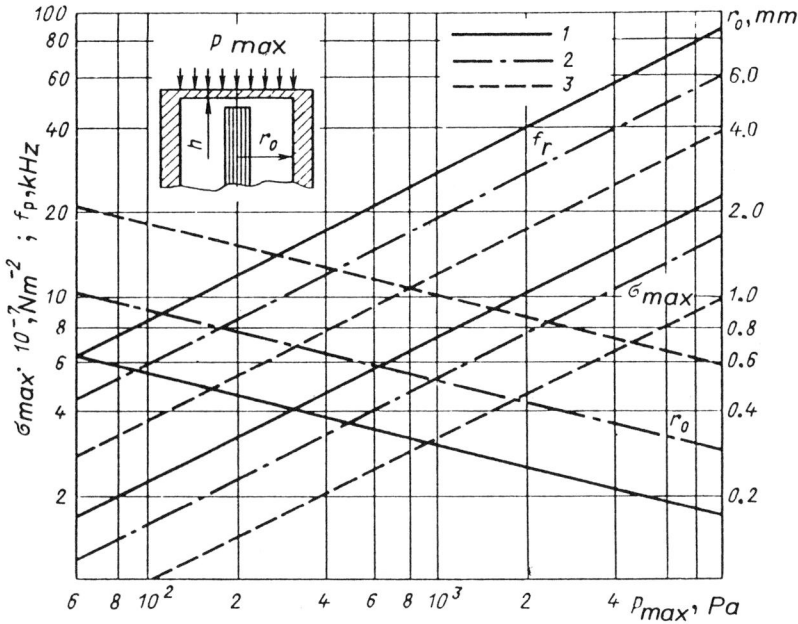

Fig.20 Optical—fiber pressure transducers with glass membranes: h, mm: $1 - 0.001$, $2 - 0.002$, $3 - 0.005$

that the membrane is rigidly clamped over its periphery on an undeformable base. Moreover, these calculations were performed on the assumption of small deflections, which is not valid here. However, tests have shown that the glass optical-fiber pressure transducers are highly effective as predicted by the calculations.

7. FABRICATION OF MINIATURE PRESSURE TRANSDUCERS AND THE RESULTS OF TESTS UPON THEM

The parameters of optical-fiber pressure transducers are largely determined by the properties of their sensitive elements. Thin glass membranes can be relatively simply prepared as follows (Fig.21).

A thin-walled glass capillary 1, sealed at one end, is locally heated by a miniature platinum heater 2 (or by a gas flame), and a bulb is inflated (Fig.21a). Because the temperature distribution during this process is nonuniform, the bulb wall thickness is not constant. A flat heater 3 is then brought up to the bulb to soften the glass, reduce the wall thickness, and produce a more or less uniform glass membrane (Fig.21b). Unless the intention is to reduce the diameter of the membrane, the process can be terminated at this stage. The next step is then to bring up a bundle of illuminating and receiving optical-fibers and fix it on the central portion of the membrane. The pressure transducer is then ready.

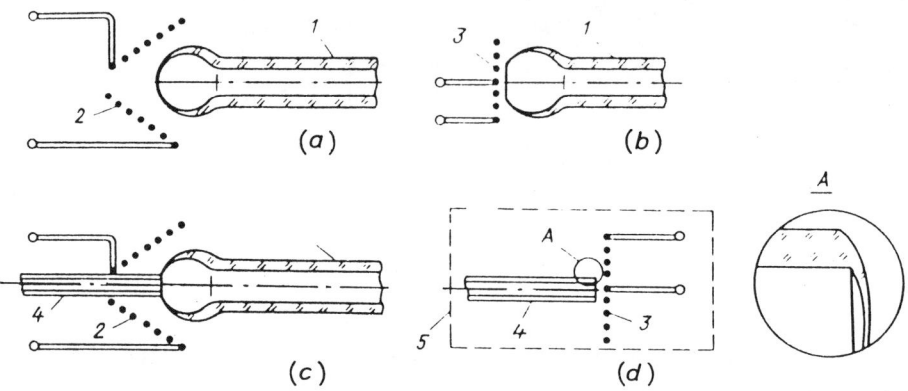

Fig.21 Fabrication of thin glass membranes

On the other hand, when it is necessary to reduce the transverse dimension of the transducer, the end of a glass capillary 4 (Fig.21c) is fused to the bulb and the original capillary 1 is removed. When the heater is turned off, the thin membrane cools more rapidly then the massive capillary, so that a nonuniform thermal stress distribution is produced, and corrugations appear on the surface of the membrane. A further operation designed to smooth out these corrugations must therefore be performed (Fig. 21d). The capillary 4 with the membrane fused to it is placed together with the flat heater 3 in a thermostat 5 held at a temperature just below the softening point of glass (about 450 °C). The power supplied to the heater 3 is then gradually increased, and the corrugations are smoothed out under visual control. It is possible, for example, to view the surface through the telescope of a cathetometer. The temperature in the thermostat is then gradually reduced to room temperature (over a period of a few hours). The membrane and capillary are thus cooled without producing any thermal stresses, so that the corrugations are not formed, and the result is a thin smooth membrane.

The membrane thickness and radial uniformity can be judged from the interference pattern produced when the surface is viewed in reflected white light. Under these conditions, one can clearly see three or four dark concentric Newton's rings, and it is well known [7] that such rings can be seen in white light only when the thickness of the membrane is equal to a few half-wavelengths of light. The characteristic half-wavelength of light is approximately 0.3 μm. Newton's rings are fringes of equal thickness. Two neighboring rings correspond to thicknesses differing by half a wavelength. Hence, it may be concluded that the minimum thickness of the membrane does not exceed 1 - 2 μm. The radial thickness variation is of the same order.

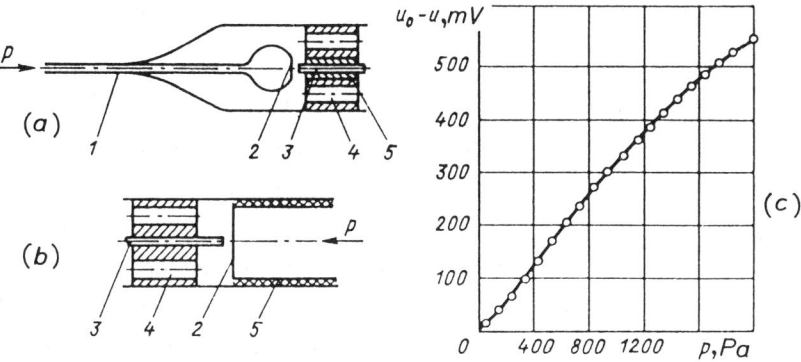

Fig.22 Optical–fiber pressure transducer (a, b) and the calibration curve for one of them (c): *1* — pulse line, *2* — membrane, *3* — fiber bundle, *4* — draining apertures, *5* — layer of epoxy compound

This technology has been used to make and test several pressure transducers used as pressure detectors in Pitot tubes. The two basic designs are illustrated in Fig.22. They can be modified by introducing the fiber bundle not as shown in Fig.22, but on the other side. Such pressure transducers are convenient for measuring the pressure on the surface of the body. The device has particularly good dynamic characteristics (see Section 16).

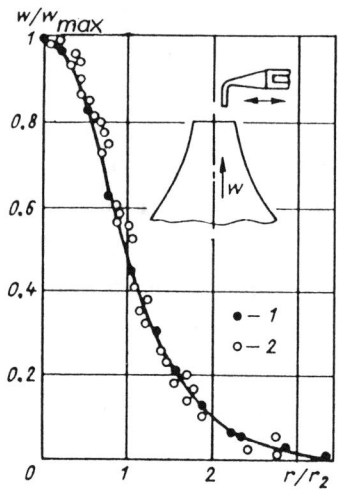

Fig.23 Measured velocity profile in a turbulent jet of nitrogen: 1 — with optical–fiber transducer, 2 — data from [6]

The most delicate operation is the mounting of the fiber bundle in the required position relative to the membrane. This can be done with a micromanipulator capable of displacing the transducer components with precision of the order of 1 μm. The components are then fixed in position by an epoxy resin, a thin layer of which does not impede adjustment, and, after polymerization, sets the components in the required positions. Special tests have shown that pressure transducers made in this way retain their characteristic on heating up to 90 °C.

Figure 22c shows the calibration curve for one of the pressure transducers of the form illustrated in Fig.22a. The diameter of the membrane two of this transducer was $2r_0 \simeq 1.4$ mm, and the maximum diameter of the entire device was 1.6 mm. The light source was a miniature lamp, and a photomultiplier was used as the detector. The pressure was determined with a micromanometer. As can be seen, the transducer is highly sensitive in the pressure range up to 2000 Pa.

A transducer with a membrane diameter of 1.8 mm, similar to that shown in Fig.21b, was employed as the pressure sensor in a Pitot tube used to measure the radial pressure distribution in an axially symmetric turbulent jet of nitrogen in stationary air. After preliminary calibration of the transducer, the system was used to measure the flow velocity at different distances from the nozzle [38]. The results are shown in Fig. 23, in which they are compared with the data reported in [6]. The usual self-similar variables are employed, namely, the ratio of the local velocity to the maximum velocity on the jet axis is plotted as a function of the ratio of the running radius r to the radius r_2 at which the velocity is lower by a factor of two as compared with its maximum value. The pressure head used in these experiments was varied in the range 0.01 – 200 Pa, indicating the high sensitivity of the optical-fiber pressure transducer.

CHAPTER
THREE

OPTICAL–FIBER VELOCITY TRANSDUCERS

8. DESIGN OF SENSITIVE ELEMENT

The sensitive element of a miniaturized velocity transducer is usually in the form of a circular cylinder placed across the flow, because the simplest procedure that can be applied to any material is to draw out a thin wire or circular fiber (capillary). Moreover, when the two perpendicular velocity components are measured, it is desirable to ensure constant sensitivity for all directions of the velocity vector, because this simplifies the process of measurement. Here, we have complete analogy with the hot–wire anemometer incorporating two crossed wires whose velocity sensitivity is made as equal as possible.

Flow past a circular cylinder. The flow across a circular cylinder and, consequently, its interaction with the cylinder is determined by the Reynolds number Re_d (see Introduction). The following flow regimes can be distinguished.

1. $Re_d \ll 1$. This is very similar to Stokes flow past a sphere.

However, in the case of the cylinder, we cannot neglect inertial forces as compared with viscous forces for arbitrarily small values of Re_d (this is the Stokes paradox). There is therefore no rigorous analytic solution of this problem. The drag coefficient can be determined from the Oseen–Lamb formula (10)

$$C_d = \frac{8\pi}{Re_d \ln(7.406/Re_d)} \tag{11}$$

which has been verified experimentally (Fig.24) for $Re_d < 0.25$.

2. $2.5 < Re_d < 50$. The boundary layer becomes detached for $Re_d \simeq 5$ and two stable vortices are formed in the separation region down the flow and are followed by a disturbed wavy wake (wholly laminar).

3. $Re_d > 50$. Here, the waves have increasing amplitude and shrink to discrete vortices. The Karman vortex sheet is established behind the cylinder. The vortex separation frequency f, and the frequency of oscillations in the lift force, which is equal to it, are given by [60]

$$fd/w \simeq 0.2\,(1 - 20/Re_d) \tag{12}$$

The amplitude of the oscillations in the lift is then comparable with the head drag. Consequently, transducers incorporating a cylindrical sensitive element cannot be used to measure turbulent velocity pulsations for $w > 50\,\nu_f/d$ where ν_f is the kinematic viscosity of the fluid. This is one of the limitations of mechanical transducers. Measurements of the time-averaged velocity are possible even when $Re_d > 50$ provided only the frequency given by (12) is not close to the resonance frequency of the sensitive element.

The drag coefficient C_d can also be calculated from the following equations that fit experimental data to within about $\pm 4\%$ (Fig. 24):

$$0.1 < Re < 1.6 \qquad C_d = 10/Re_d^{0.77} \tag{13}$$

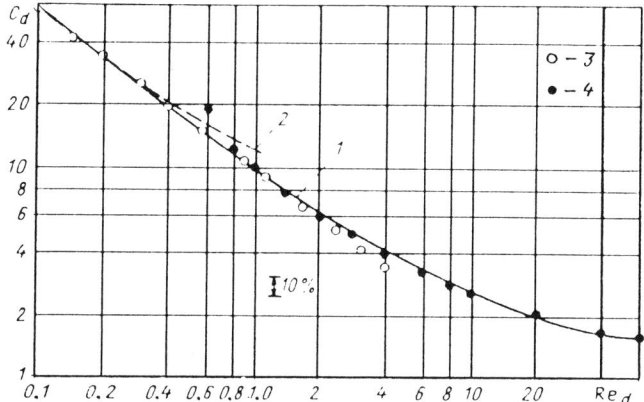

Fig.24 Drag coefficient of circular cylinders placed transversely to flow: *1* — experimental curves; calculations based on (*11*), (*13*), and (*14*) — *2*, *3*, and *4*, respectively

$$0.8 < \text{Re}_d < 60 \qquad C_d = 5.06/(\log \text{Re}_d + 0.6)^{1.36} \qquad (14)$$

The head drag per unit length of the cylinder placed transversely in the flow is given by

$$q = C_d \rho_f w^2 d/2 \qquad (15)$$

This force is a function of the flow velocity w and gives rise to the deflection of the sensitive element which is recorded by the optical-fiber displacement transducer.

Validity of the quasi–stationary approach. Since, so far, it has not been possible to develop standard sources of time-dependent velocity, turbulent velocity fluctuations are normally measured using transducers calibrated under static conditions. It is assumed that the instantaneous signal produced by the transducer at any given time is related to the instantaneous velocity given by the calibration curve, i.e., the measurement process is assumed to be quasi-stationary. It is clear that this is possible only when the characteristic time for the development of the boundary layer on the sensitive element is significantly shorter than the minimum period of the velocity fluctuations to be measured.

The time for the development of the boundary layer can be estimated from the expression recommended in [60]. This gives the time τ_{sep} for the boundary layer to start separating from a cylinder that is abruptly set in motion with constant velocity w:

$$\tau_{sep} = d/[4w(1 + 4\pi/3)]$$

For example, when $d = 0.02$ mm and $w = 1$ m/s, we have $\tau_{sep} \simeq 10^{-6}$ s. This shows that the flow is established very rapidly, and that the static calibration curve can indeed be used to measure velocity fluctuations.

Design of sensitive element. A velocity transducer designed for measurements in time-dependent flows must satisfy two mutually exclusive requirements, namely, high sensitivity (i.e., low stiffness of sensitive element) and high resonance (natural) frequency of the sensitive element (i.e., high stiffness). The latter is necessary because the resonance frequency determines the upper limit of frequency measurement, since the dynamic distortion of the measured quantity increases rapidly as the resonance frequency is approached.

One of the basic principles used in the design of mechanical velocity transducers is therefore the requirement that the sensitive element must have the maximum possible frequency for given transducer sensitivity. The latter is determined from the required precision of measurement.

Consider the composite optical-fiber velocity transducer. The first transducer (sensitive element) transforms the flow velocity w into the deflection Δr of the sensitive element, so that, using (15), we can write its transfer function in the form [41]

$$\Delta r = ql^4/nEI \sim C_d w^2 dl^4/nEI$$

where d and l are, respectively, the diameter and length of the sensitive element, n is a factor that depends on the design of the sensitive element (the number and types of support), E is Young's modulus of the material of the sensitive element, I is the second moment of area of the cross section of the sensitive element, and C_d is given by (11), (13), or (14), depending on the value of Re_d.

The second transducer (optical–fiber displacement transducer) transforms Δr into a change $\Delta \Phi = \Phi - \Phi_0$ in the light flux, where Φ_0 is the value of Φ at $\Delta r = 0$. If we use a transducer that measures the displacement of an opaque body (see Section 3) on the linear portion of its characteristic (see Figs. 10b and 11b), then

$$\Delta \Phi \sim r_r \Delta r \ .$$

The third transducer (photodetector) transforms $\Delta \Phi$ into an increase in the output electrical signal $\Delta u = u - u_0$, where

$$\Delta u \sim \Delta \Phi$$

Hence,

$$\Delta u \sim r_r \Delta r \sim C_d w^2 r_r dl^4 / (nEI)$$

The sensitivity of the transducer as a whole can be defined by

$$\frac{\partial (\Delta u)}{\partial w} \sim \frac{r_r dl^4}{nEI} \frac{\partial}{\partial w} (C_d w^2)$$

and, within the framework of this analysis, the senstivity can be regarded as set by the required precision of velocity measurement.

The characteristic of the displacement transducer is single–valued if $d \geqslant 2r_r$. It is best to have $d \sim 2r_r$ because this produces improved spatial resolution and the limitation $Re_d < 50$ has little effect up to the higher values of w. The result is that

$$l \sim \left(\frac{nEI}{r_r^2}\right)^{1/4} \left[\frac{\partial}{\partial w}(C_d w^2)\right]^{-1/4}$$

The resonance (natural) frequency of the sensitive element (see Section 14) is given by

$$f_r = \frac{r_1^2}{2\pi l^2}\left(\frac{EI}{\rho_{se} F_{se}}\right)^{1/2} \sim \left(\frac{r_1^2}{n^{1/2}}\frac{1}{\rho_{se}^{1/2}}\right)\left[\frac{\partial}{\partial w}(w^2 C_d)\right]^{1/2} \qquad (16)$$

where F_{se} is the cross sectional area of the sensitive element, ρ_{se} is the density of its material, and r_1 is a coefficient that depends on the design of the sensitive element.

This expression enables us to select the optimum design of the sensitive element (the values of r_1 and n) and its material (the value of ρ_{se}), and also to draw certain conclusions about the desirable value of d (or r_r).

As far as the configuration of the sensitive element is concerned, we need only consider rigidly-clamped beams because experiment shows that the presence of hinged supports produces undesirable frictional forces. A suitable design is therefore one of the following two: a cantilever, rigidly clamped at one end, with $r_1 = 1.875$, $n = 8$, $r_1^2 n^{-1/2} = 1.243$, or a single-span beam, rigidly clamped at both ends, with $r_1 = 4.730$, $n = 384$, i.e., $r_1^2 n^{-1/2} = 1.142$ (see [41] and [57]). Consequently, other things being equal, the cantilever configuration has a resonance frequency that is higher by about 10% as compared with the beam clamped at both ends. The cantilever design is also preferable from the point of view of compactness of the velocity transducer, and can be satisfactorily combined with the optical-fiber displacement transducer.

It also follows from (16) that

$$f_r \sim \rho_{se}^{-1/2}$$

so that low-density materials are preferable. As noted in Section 5, glass is the best material from this point of view (see Table 1). The resonance frequency of the glass sensitive element is lower by a factor of almost two than, for example, the comparable figure for a steel element. Moreover, glass has the additional advantage that its Young's modulus is relatively independent of temperature (see Fig.19).

It can be shown that, in the first approximation,

$$\left[\frac{\partial}{\partial w}(w^2 C_d)\right]^{1/2} \sim w^m/d^{1-m}$$

where m varies from 0.12 for $Re_d < 1$ to 0.33 for $Re_d \simeq 50$. Consequently, a reduction in the diameter of the sensitive element produces an increase in f_r. In addition, f_r increases with increasing characteristic flow rate w, which corresponds to the well known phenomenon observed in turbulent flows, namely, the fact that the upper limit of the flow spectrum shifts towards higher frequencies as the flow rate increases.

Another attractive feature of mechanical velocity transducers is that they have no lower limit: since the deflection is $\Delta r \sim l^4$, very low flow rates can be measured by increasing l.

The above analysis was given for the optical-fiber transducer of the displacement of an opaque body. However, the results are also valid when the deflection of the sensitive element is recorded by a transducer of the displacement of the illuminating lightguide (see Section 4).

9. OPTICAL—FIBER TRANSDUCERS OF THE VELOCITY OF A TRANSPARENT FLUID

Optical-fiber displacement transducers enable us to miniaturize mechanical velocity transducers and, as shown in Section 8, to produce a radical improvement in their metrological properties. In the case of transparent media, the displacement transducer can work directly in the working fluid, and its velocity calibration can be performed in the same fluid. The point is that, in addition to viscosity and density, the calibration curve may depend on the refractive index of the fluid. Whilst density and viscosity effects can be taken into account relatively simply, using well known similarity relations, the effect of the refractive index is more difficult to allow for because the angular distribution of radiation at the end of the illuminating lightguide is not usually known.

Velocity transducers using the displacement of an opaque body. When the displacement transducer relies on the displacement of an opaque body (see Section 3), the rays of light employed in the measurement process lie in the plane perpendicular to the axis of the sensitive element. However, in order to minimize the transverse dimensions of the probe introduced into the flow, it is desirable to have the illuminating and receiving lightguides parallel to the axis of the sensitive element, with the sources of light and the photodetectors located outside the flow. There are two designs of optical-fiber displacement transducers based on these ideas (Fig.25).

In the first design, (Fig. 25a), the transducer consists of lightguides 0.12 - 0.14 mm in diameter, whose ends are cut and polished at 45° to the lightguide axis. The regularly-packed fiber bundle consists of several hundred fibers can then be arranged to fit closely in a glass tube and are bonded with resin. The end of the tube containing the lightguides is then cut at 45° to the tube axis, and the resulting surface is polished and the resin removed with a suitable solvent. The blanks produced in this way are then successively coated with aluminum, copper, and zinc black. The coatings are produced by evaporation in vacuum. A sharp blade is then used to remove the metal coatings near the polished ends until a gap 0.15 mm long and 0.015 - 0.020 mm wide is produced. A quartet of such lightguides (illuminating *1* and receiving *2*) is then assembled in the form of a two-component displacement transducer. The crossed beams used in this transducer are produced by reflection from the metal coating of the ends

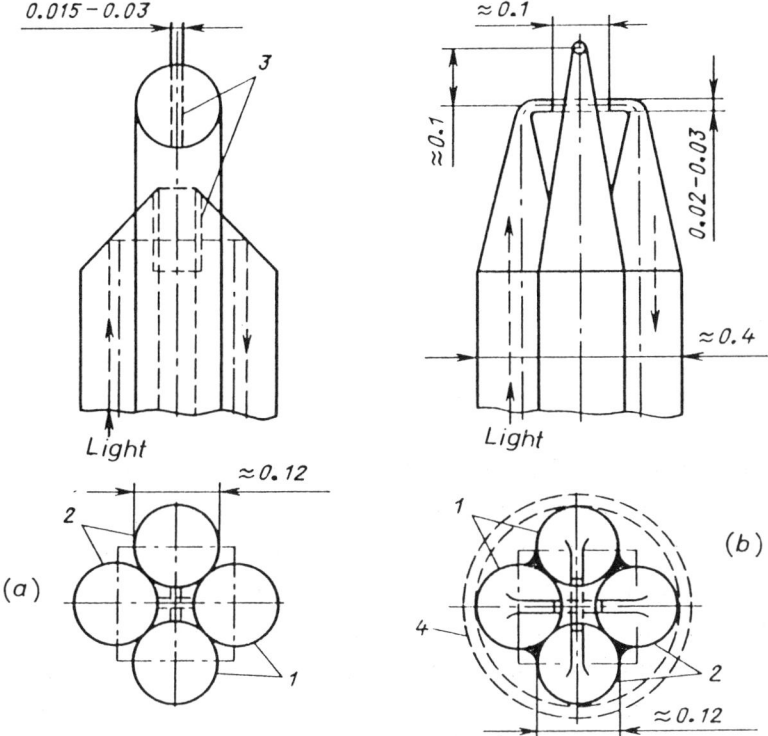

Fig.25 Optical—fiber displacement transducers

and are transmitted by slits running along the opaque coating. The advantage of this design is the relatively high intensity reaching the photodetector and the fact that the signal is a linear function of displacement. Nevertheless, these transducers have had limited range of application because they are difficult to produce and to adjust. It is also difficult to ensure that the measuring rays are made perpendicular during assembly because there are no visual direction indicators.

The two-component displacement transducer shown in Fig. 25b is more widely used. Here the two crossed beams are produced by two pairs of bent lightguides.

The quartet of lightguides of roughly the same diameter is inserted into a copper capillary 2 whose diameter is arranged so that the lightguides

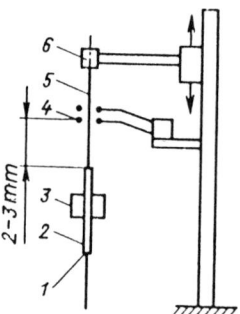

Fig.26 Principle of a device for pulling lightguides

occupy the position shown in Fig. 25b. The quartet, which is first glued together, is then inserted into the device shown in Fig. 26 (the lightguides are glued together at 1). The lightguides 5 are held in the clamp 6 and are centered inside the platinum heater 4 consisting of two turns of wire, 0.3 mm in diameter. The heater current and the weight of the load 3 fixed to the copper capillary 2 are chosen so that the length of the region over which the diameter of the lightguides 5 is reduced from the initial value down to 0.015 mm is about 2 mm. The lightguides are then cut off at a distance of about 5 mm from the beginning of this segment and are inserted into a micromanipulator in which the optical system of the transducer is fabricated. The lightguides are bent, cut, and their ends are lightly fixed until they assume the configuration shown in Fig. 25b. Each pair of lightguides is adjusted separately. The illuminating lightguide is coupled to the light source and the receiving lightguide to the photodetector. The tips of the lightguides continue to be heated by platinum microheaters until they bend so that the signal from the corresponding detector is a maximum.

Both designs of displacement transducers are used in velocity transducers for transparent bodies.

Figure 27a shows a velocity transducer incorporating the displacement transducer of Fig. 25a. The lightguides 2 are glued to a glass tube 3. The sensitive element 1 is obtained by pulling it out of the glass capillary 4 and, like the lightguides, is coated with aluminum and zinc

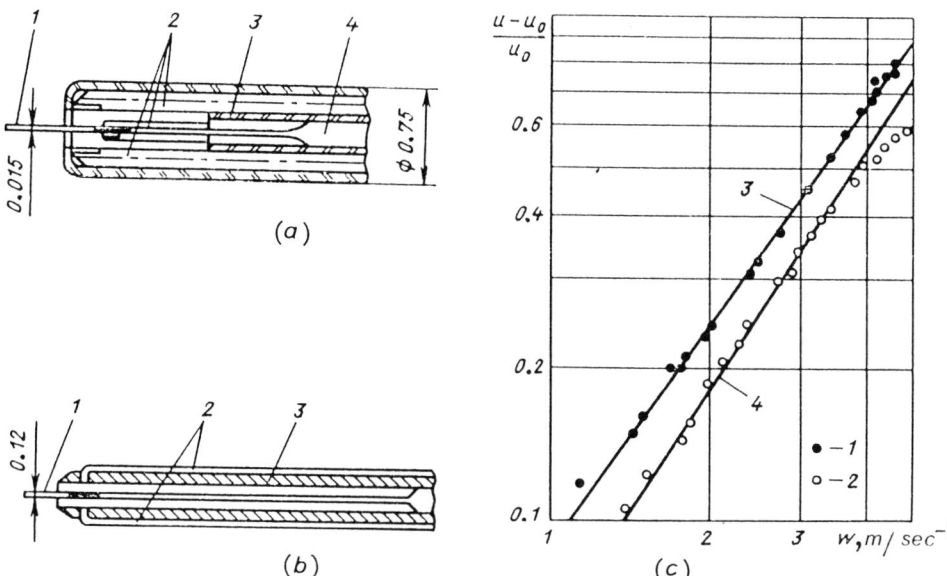

Fig.27 Optical–fiber velocity transducers incorporating lightguides with reflecting ends (a) and bent lightguides (b), together with their calibration curves (c): *1* — transducer shown in Fig.27a, *2* — transducer shown in Fig. 27b, *3, 4* — calculated from (14), (15), and (17)

black, deposited by evaporation in vacuum. The presence of the electrically conducting coatings on the sensitive element and on the lightguides means that variable electrical potentials can be applied to them and the velocity transducer can be calibrated dynamically (see Section 14).

The signal generated by this transducer is shown in Fig. 27c (full points) as a function of the flow rate. Because the characteristic of the displacement transducer used in this device is linear, its signal can be calculated from the formula

$$\bar{u} = (u - u_0)/u_0 = kC_d w^2 \qquad (17)$$

where k is a constant.

Estimates of the Reynolds number Re_d show that (14) must be used to calculate C_d. Figure 27c also shows the values calculated from (14), (15), and (17). They can be made to agree with experiment when the constant k is suitably chosen. It is clear that for $w = 1.5 - 5$ m/s the calibration curve can be satisfactorily represented by the power-type formula

$$\overline{u} = cw^n \qquad (18)$$

Velocity transducers based on the second design (see Fig. 25b) of the displacement transducer for an opaque body [2, 37] have also been made and tested in air flows. The design of one such transducer is illustrated in Fig. 27b. The deflection of the glass sensitive element 1 is recorded by the two bent lightguides 2, fixed to the metal shield 3. The calibration graph for this transducer in a flow of air is shown in Fig. 27c (open circles) together with the curve calculated from (14), (15), and (17). Here again, the calculated values are in satisfactory agreement with experiment in the range $w = 1.4 - 4$ m/s, indicating that the displacement transducer is linear, and confirming the conclusions based on the results obtained with the displacement transducers for opaque bodies (see Section 3). The power-type approximation (18) can also be used in this range.

The above properties of the calibration curves can be used to set up a relatively simple algorithm for determining the two velocity components w_1 and w_2 from the measured signals $\overline{u}_1$ and $\overline{u}_2$ produced by optical-fber displacement transducers lying along perpendicular axes x_1 and x_2 (the vector w is assumed to lie in the plane perpendicular to the axis of the sensitive element).

The displacement vector Δr is always parallel to the velocity of vector w, where

$$\Delta r = c_1 |w| w C_d$$

$$\frac{\Delta r_1}{|\Delta r|} = \frac{w_1}{|w|}$$

$$\frac{\Delta r_2}{|\Delta r|} = \frac{w_2}{|w|}$$

Consequently,

$$\overline{u_1} = c_2 \Delta r_1 = c_3 C_d |w| w_1 \tag{19}$$

$$\overline{u_2} = c_3 \Delta r_2 = c_5 C_d |w| w_2 \tag{20}$$

We then have $|w| = (w_1^2 + w_2^2)^{1/2} = w_1(1 + w_2^2/w_1^2)^{1/2}$ and, when the ratio w_2/w_1 is small (which is usually the case in the turbulent flows), we can assume that $|w| \simeq w_1$. When $w_2/w_1 = 0.3$, this equation is satisfied to within about 4%. Instead of (19) and (20), we then have

$$\overline{u_1} \simeq c_3 C_d w_1^2 \tag{21}$$

$$\overline{u_2} \simeq c_5 C_d w_1 w_2 = c_6 \overline{u_1} w_2/w_1 \tag{22}$$

If, at the same time, (18) is satisfied, then

$$\overline{u_1} = c_7 w_1^n \tag{23}$$

$$\overline{u_2} = c_8 w_1^{n-1} w_2 = c_9 w_2 \overline{u_1}/w_1 \tag{24}$$

Hence, having determined c_7, c_9, and n by calibrating the transducer, and having measured $\overline{u_1}$ and $\overline{u_2}$, we can simply determine w_1 and w_2. Moreover, using (23), we can design a simple functional linearizer for the velocity transducer in which an electronic circuit is used to raise the output quantity to the necessary power (see Section 19).

The above velocity transducers employ shields that have relatively large transverse dimensions (about 1 mm), so that their use is usually confined to measurements of velocity and its turbulent fluctuations well away from channel walls, or in jet flows, and so on. The measurements are carried out in relatively small flow-velocity gradients, i.e., in relatively small velocity ranges. The above calibration properties of the transducers can then be used, and the measuring equipment can be substantially simplified.

Velocity transducers using the displacement of an illuminating lightguide. When the velocity is measured near the walls of a channel or the surface of a body in a flow, it is more convenient to employ optical-fiber velocity transducers that employ the displacement of the illuminating lightguide [3, 14] (see Section 4). These are relatively simple devices. For example, Fig. 28 shows a simple transducer that is nevertheless very sensitive to velocity. The illuminating *2* and receiving *1* lightguides are held against the surface of a copper wedge *3* by a bronze spring *4*. The sensitive element whose deflection is used to determine the velocity of the incident flow is a segment of a lightguide, 0.025 in diameter and 4 mm long.

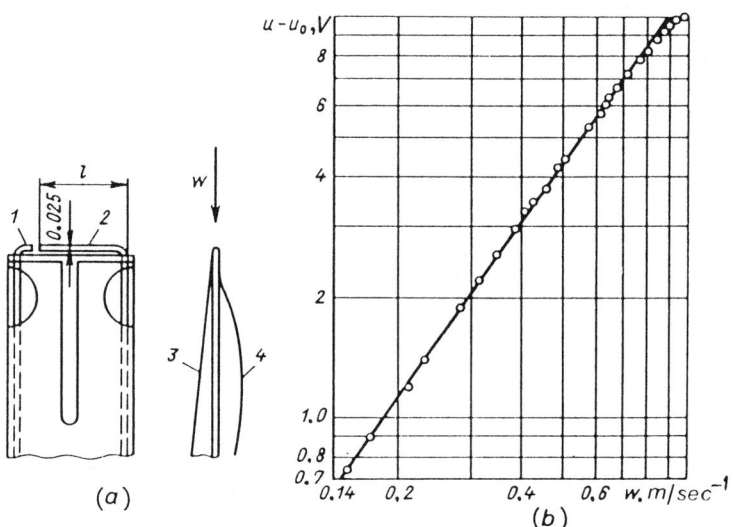

Fig.28 Principle of a simple velocity transducer (*a*) and its calibration curve (*b*)

The dependence of the output signal $\Delta u = u - u_0$ of this transducer on the air flow rate is shown in Fig. 28*b*. As can be seen, the photomultiplier signal changes by 10 V as the flow rate is varied between zero and 1 m/s.

The calibration curve is satisfactorily reproduced and is again described by a power law such as that given by (23). This design is satisfactory only under isothermal conditions, i.e., when the adjustment, calibration, and measurements are performed at the same temperature. Unless this is so, measurements are impossible because materials with different thermal expansion coefficients are employed and the characteristics of the transducers depend significantly on temperature.

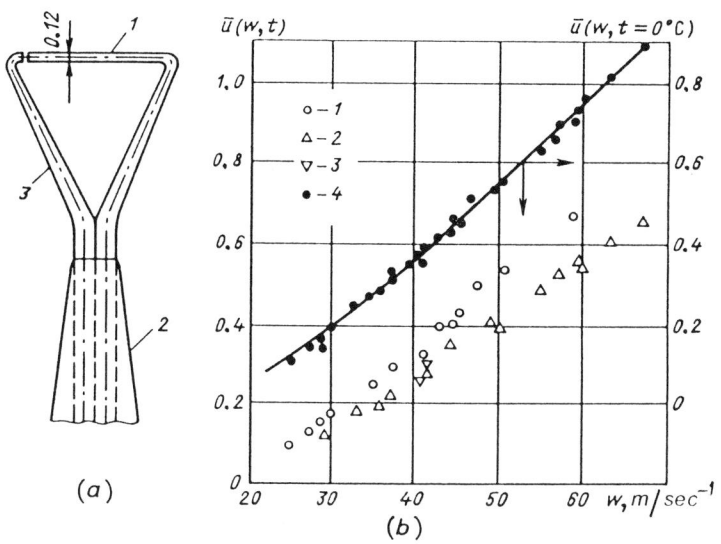

Fig.29 Glass velocity transducer (a) and its calibration in a flow of hot air (b): temperature, °C: $1 - 15 - 30$, $2 - 50 - 100$, $3 - 100 - 150$, $4 -$ after correction for the change in the properties of air

The only way this can be avoided is by using an all-glass design (Fig. 29a). The lightguides in this configuration (illuminating 1 and receiving 3 lightguides) are fused into the capillary holder 2 made from glass with a lower softening point than the lightguide material. This transducer is calibrated in a stream of hot air, and results are as shown in Fig. 29b. It is clear from the figure that the calibration curve $\bar{u}(w)$ is not single-valued, i.e., measurements made at different temperatures between 20 and 150°C (open circles) do not lie on the same curve.

The two basic reasons for this phenomenon are, first, the deformation of the transducer elements by thermal stresses, and, second, the viscosity and density of air that appear in (15) are functions of temperature. The second factor can be excluded by introducing the necessary correction due to the temperature dependence of the density ρ_f and viscosity ν_f of air. Under our experimental conditions,

$$C_d \simeq 1/\mathrm{Re}_d^{0.12} \simeq \nu_f^{0.12}$$

Conseqently, the correction for the change in the properties of air can be introduced by multipling $\overline{u}$ by the factor

$$k_t = \left(\rho_f \nu_f^{0.12}\right)_{t=0°\mathrm{C}} \Big/ \left(\rho_f \nu_f^{0.12}\right)_t$$

When the measurements are normalized to $t = 0$ °C, the experimental points are found to lie on a single curve (full points in Fig. 29b). Consequently, after calibration at some known temperature, a transducer of this kind can be used in hot gas streams. The temperature at which the velocity is measured must be known, and the appropriate correction must be introduced into the measurements.

As noted in Section 4, transducers that measure the displacement of the illuminating lightguide can be used to determine the two displacement components and, hence, the velocity. This possibility has been verified with the two-component optical-fiber velocity transducers [14] shown in Fig. 30a. The illuminating lightguide 2, 0.14 mm in diameter, is drawn out down to a diameter of 0.008 mm and is bent as shown in the Figure. Its end is arranged to lie against the two ends of the receiving lightguide 4, which were also made from 0.14-mm diameter lightguides drawn out to a diameter of 0.016 mm. The thin part of the illuminating lightguide (1.1 mm long) is the sensitive element 1. All three lightguides are fused into the glass holder 3 and are fixed in a measuring probe.

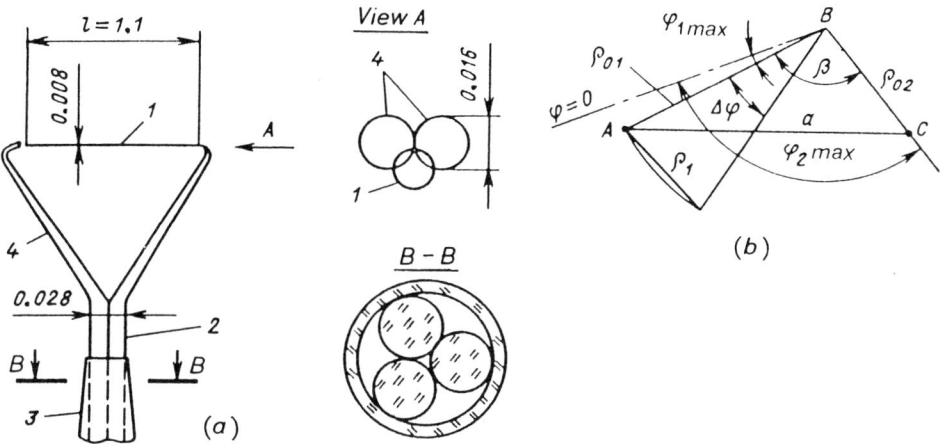

Fig.30 Two-component velocity transducer (a) and diagram illustrating the calibration process (b)

Measurements of the two velocity components with this transducer are preceded by a relatively complicated calibration. The transducer is placed in a standardizing device capable of altering the magnitude and direction of the velocity of air at right angles to the axis of the sensitive element. The latter is placed in the nozzle of the standardizing device in the region with a uniform velocity profile. The transducer can be rotated through an angle ϕ around the axis of the sensitive element. In the initial state, the axis of the receiving lightguide (Fig. 30b) cuts the plane of the figure at points A and C, and the axis of the sensitive element cuts this plane at B. The two components of the displacement of the end of the sensitive element (and of the velocity) can be measured only when the signals u_1 and u_2 from the photodetectors coupled to the receiving lightguides depend only on the separations ρ_1 and ρ_2 between the axis of the central elements and the axis of the receiving lightguides, respectively. In other words, the lines of equal illumination in the plane of Fig. 30b must be concentric circles with centers lying on the axis of the sensitive element. This means that symmetrization of the light beams is a useful property of these lightguides (see Section 1). The axial symmetry of the end of this sensitive element also plays an important part, so that it is necessary to fire the end lightly in order to remove surface irregularities produced when the lightguide is cut.

The calibration of the two-component velocity transducer is based on the assumption that the illumination is symmetric relative to the axis of the sensitive element. The first step is to find the angle ϕ_{1max} and velocity w_{1max} for which the signal u_1 reaches its maximum value u_{1max}, i.e., $\overline{u}_1 = u_1/u_{1max} = 1$. This means that the end of the sensitive element is displaced by the distance ρ_{01} to the point A. Next, holding the value of w constant and equal to $2w_{1max}$, we vary the angle ϕ and obtain the function $\overline{u}_1(\Delta\phi)$ shown in Fig.31a. Since

$$\rho_1 = 2\rho_{01} \sin \frac{1}{2} \Delta\phi$$

and

$$\overline{\rho}_1 = \rho_1/\rho_{01} = 2 \sin \frac{1}{2} \Delta\phi$$

the function $\overline{u}_1(\Delta\phi)$ can be used to determine $\overline{u}_1(\overline{\rho}_1)$ (Fig. 31b). The function $\overline{u}_1(w)$ of Fig. 31c is then obtained for $\phi = \phi_{1max} = $ const, and this is then used together with $\overline{u}_1(\overline{\rho}_1)$ obtained previously to determine $\rho_1(w)$ (Fig. 31d). The displacement Δr of the sensitive element due to a velocity w is determined from the relation $\Delta r = \rho_{01} - \rho_1$ or $\overline{\Delta r}_1 = \Delta r/\rho_{01} = 1 - \overline{\rho}_1$ and the known functions $\overline{u}_1(\overline{\rho}_1)$ and $\overline{u}_1(w)$. A typical form of $\overline{\Delta r}_1(w)$ is shown Fig. 31e.

The entire procedure is then repeated for the second receiving lightguide. A determination is made of ϕ_{2max}, w_{2max}, the function $u_2(\rho_2)$ and finally, the quantity

$$\overline{\Delta r}_2 = \Delta r/\rho_{02} = \overline{\Delta r}_2(w)$$

(Fig. 31f). It is clear that the ratio

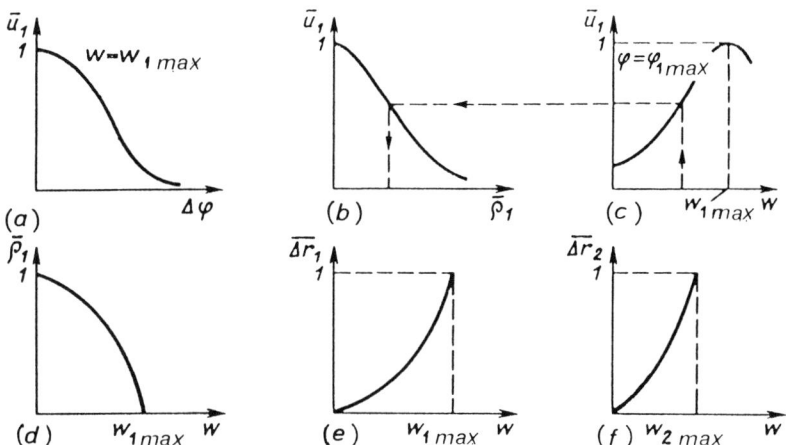

Fig.31 Calibration of a two-component velocity transducer

$$\overline{\Delta r_1}/\overline{\Delta r_2} = \rho_{02}/\rho_{01} = \overline{\rho}_{02}$$

should be constant (independent of velocity w) if the assumption of symmetric illumination, upon which the calibration procedure is based, is indeed correct. Experiments confirm this.

Finally, the dimensionless separation $\overline{a} = a/\rho_{01}$ between the axes of the receiving lightguides must be calculated. It is clear from Fig. 30b that

$$a^2 = \rho_{01}^2 + \rho_{02}^2 - 2\rho_{01}\rho_{02}\cos\beta$$

and

$$\overline{a}_2^{-2} = a^2/\rho_{01}^2 = 1 + \overline{\rho}_{02}^2 - 2\overline{\rho}_{02}\cos\beta$$

where $\beta = \phi_{2\,max} - \phi_{1\,max}$.

This method is to determine $\rho_{01} = 1$, $\overline{\rho}_{02}$, $\overline{a}$, and the functions $\overline{u}_1(\rho_1)$ and $\overline{u}_2(\rho_2)$ that are necessary to find the two velocity components from the measured $\overline{u}_1$ and $\overline{u}_2$.

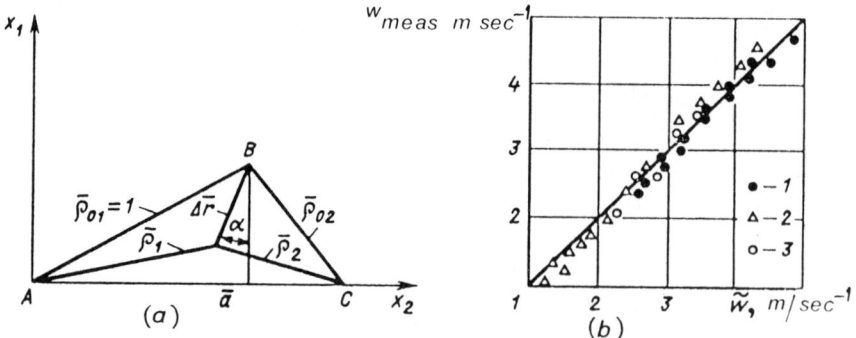

Fig.32 Diagrams illustrating the measurement of velocity (a) and comparison between measured and specified values (b): $1 - |w|$, $2 - w_1$, $3 - w_2$

Suppose that the orthogonal coordinate axes x_1 and x_2 are oriented as shown in Fig. 32a. By measuring $\bar{u}_1$ and $\bar{u}_2$ we obtain $\bar{\rho}_1$ and $\bar{\rho}_2$ and the direction and magnitude of w. The components w_1 and w_2 of w are given by $w_1 = |w| \sin \alpha$, $w_2 = |w| \cos \alpha$. The standardizing device used to calibrate the transducer is also used to verify the validity of this method of measuring $|w|$, w_1, and w_2. The values of $|w|$, w_1, and w_2 are specified by the standardizing devices and were then determined from the measured $\bar{u}_1$ and $\bar{u}_2$. The results can be judged from Fig. 32b where the values of $|w|$, w_1, and w_2 are plotted along the vertical axis and the given values of these quantities along the horizontal axis. The measured and given values agree to within a root-mean-square uncertainty of about 5% (the maximum deviation being 15%). Analysis of the uncertainties shows that the main reason for the measurement uncertainties is the photomultiplier instability. This becomes quite clear in the course of prolonged calibration (up to a few hours). Steps were therefore taken to stabilize the characteristics of the light-intensity measuring system (see Chapter 5).

Spatial resolution. The optical-fiber velocity transducers discussed above have high spatial resolution. When only one of the velocity components is measured, these transducers match the resolution of hot-wire anemometers. However, they are much better than the latter when two components are measured because, in contrast to the hot-wire anemometer, only one sensitive element is now necessary. This becomes the dominant factor when velocity measurements are carried out in the immediate neighborhood of the wall, since the two-component hot-wire anemometer cannot be brought closer together to the wall than, say, half the length of the wire. On the other hand, for the optical-fiber transducer, this distance amounts to roughly the radius of the sensitive element, i.e., it is smaller by two orders of magnitude. When the two velocity components are measured, only the laser Doppler anemometer [17] has a spatial resolution comparable with that of the optical-fiber transducer.

One of the sources of error in the case of the hot-wire anomometer is that the wire holders affect the output signal. The hot-wire anemometers have roughly the same velocity sensitivity along their entire length, so that the disturbance of the flow by the wire holders introduces an additional uncertainty into velocity measurements.

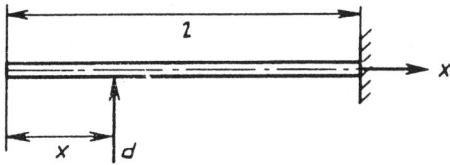

Fig.33

The optical-fiber transducer shown schematically in Fig.30 does not suffer from this defect because its velocity sensitivity varies along the length of the sensitive element: it is a maximum at its free end and zero at the clamped end. Some idea of the distribution of sensitivity along the length of the sensitive element can be obtained by considering its deflection as a function of the position of the point of application of a localised force P (Fig. 33). At the point at which the force is applied the deflection is [41]

$$\Delta r(x) = P(l-x)^3 / 3EI$$

The angle of rotation of the cross section at the point x is

$$\theta(x) = P(l-x)^2/2EI$$

and the deflection at $x = 0$ due to a force applied at x is

$$\Delta r_{z=0}(x) = \Delta r(x) + x\,\theta(x)$$

$$= \frac{P}{EI}\left[\frac{1}{3}(l-x)^3 + \frac{1}{2}x(l-x)^2\right]$$

Let $\Delta r_{z=0}(x)/\Delta r_{z=0}(0) = \psi(x)$ and let us refer to this as the influence function. Hence

$$\psi(\bar{x}) = (1-\bar{x})^3 + \frac{3}{2}\bar{x}(1-\bar{x})^2 \qquad (25)$$

where $\bar{x} = x/l$.

The function $\psi(\bar{x})$ represents the distribution of sensitivity along the length of the sensitive element. It is clear from Fig. 34 that the sensitivity is a maximum at the end of the sensitive element and falls rapidly with distance from it. For example, $\psi(\bar{x}) \simeq 0.06$ at $\bar{x} = 0.8$.

The function ψ can be used to find the deflection of the sensitive element under a load $q(x)$ distributed arbitrarily along its length. The transfer function of a transducer can then be determined from ψ when the velocity varies along the length of the sensitive element [52]. The deflection is given by

$$\Delta r_{z=0} = \left[l^4/3EI\right]\int q(\bar{x})\,\psi(\bar{x})\,d\bar{x} \qquad (26)$$

where the integral is evaluated between 0 and 1.

Figure 35 shows the experimental results obtained with the transducer shown in Fig. 30a. The signal $\bar{u}_1$ from one of the photodetectors was measured as a function of the position of a thin metal screen covering a length l_e of the sensitive element. The ratio $\bar{u}_1$ of the value of u_1 to its maximum value u_{1max} corresponding to $l_e = 0$ is plotted along the vertical axis. The figure also shows the calculated $\bar{u}_1(\bar{l}_e)$ deduced from (26):

$$\bar{u}_1 = \frac{u_1}{u_{1max}} \simeq \frac{\Delta r_{z=0,l_e}}{\Delta r_{z=0,l_{e=0}}} =$$

$$= \frac{\int \psi(\bar{x})\, d\bar{x}}{\int \psi(\bar{x})\, d\bar{x}} = 1 - \frac{4}{3}\bar{l}_e^{3} + \frac{1}{3}\bar{l}_e^{4} \qquad (27)$$

where $\bar{l}_e = l_e/l_0$, the integral in the numerator is evaluated between 0 and $1-\bar{l}_e$, and the integral in the denominator between 0 and 1.

The calculations are in satisfactory agreement with experimental results. The discrepancies for $\bar{l}_e \to 1$ can be explained by vortices separating from the edge of the screen and influencing the signal (the end region of the sensitive element has the maximum sensitivity). This is also indicated by the fact that the transducer signal becomes unstable as $\bar{l}_e \to 1$.

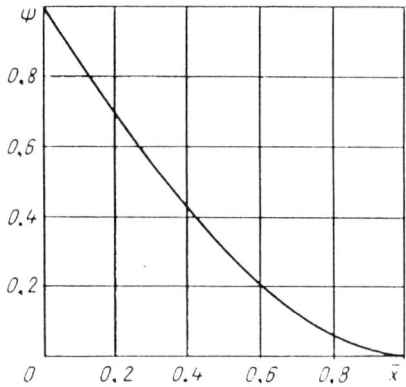

Fig.34 Graph of the influence function

10. OPTICAL–FIBER VELOCITY TRANSDUCERS FOR OPAQUE FLUIDS

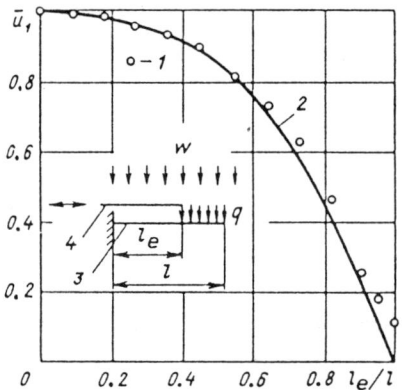

Fig.35 Effect of shielding of the sensitive element on the transducer signal: 1 — experimental points, 2 — calculated from (27), 3 — sensitive element, 4 — shield

In the optical–fiber velocity transducers described in the last Section, the rays of light pass through a transparent working fluid. On the other hand, studies of fluid flow often involve opaque fluids, including liquid metals. Important advances have been made in magnetohydrodynamics, i.e., in the study of physical phenomena associated with the interaction of electrically conducting fluids and the electromagnetic field. The parameters of this interaction are in many ways determined by the electrical conductivity of the fluid, so that many of the experiments involve the use of liquid metals. Researchers have been particularly interested in the suppression of turbulence by a magnetic field and the transformation of its spatial structure by the fields.

Recent years have seen the development of research into electromotive flow [62] which was stimulated by the needs of industry in areas such as the production of aluminum and steel slag furnaces. These processes involve very high electrical currents (some tens of kiloamperes) flowing through molten metal or slag. The interaction between the current and its own magnetic field produces intensive electromotive flows that are difficult to calculate. Liquid metals are the only media that can be used to simulate such processes under laboratory conditions.

Limitations of hot−wire anemometry and conductive anemometry. Until quite recently, experimenters have had at their disposal only two basic methods for measuring the characteristics of turbulent flows in liquid metals, namely, the hot−wire anemometer [52] and the conductive anemometer [25, 42]. There are certain basic difficulties in using these methods. They are due both to the properties of liquid metals and the properties of flows in such metals.

The operation of the hot−wire anemometer in a liquid metal is complicated by the following factors. It is well−known that a contact thermal resistance is produced on the separation boundary between the sensitive element and the liquid metal because of the deposition of oxides that are always present in the metal even after careful purification. This contact resistance is a function of time because it depends on deposition dynamics and the removal of contamination, and ultimately leads to an instability of the transfer function of the device and to significant errors in the measured velocity [31].

The deposition of an insulating layer on the hot wire produces a significant deterioration in the dynamic characteristics of the instrument because of the higher thermal inertia of the wire. However, the high thermal conductivity of liquid metals (or, more correctly, the low values of the Prandtl number) and the associated large thickness of the thermal boundary layer on the sensitive element give rise to the most unfavorable effect on the dynamic characteristics of the device. In particular, it is shown in [32] that the deposition of a glass insulator 0.004 mm thick on the 0.009 − mm diameter wire reduces the upper frequency limit to 300 Hz for high flow rates. Effects associated with the thermal boundary layer are more important at low flow rates and we reduce this limit down to 50 Hz at 10 cm/s. At the same time, it is well known [25] that the spectrum of flows with this mean velocity contains significant components at frequencies up to at least 100 Hz.

The conductive anemometer is based on the phenomenon of electromagnetic induction. The flow rate is related to the magnetic induction B and the gradient of the potential u by the following vector equation:

$$j/\sigma = - \operatorname{grad} u + w \times B$$

where j is the electric current density and σ is the electrical conductivity of the fluid.

Consider the simple case where the velocity has only one component w_1 and the magnetic induction B points along the x_2 axis which is perpendicular to the velocity. We then have

$$\partial u/\partial x_2 = - j_2/\sigma$$

$$\partial u/\partial x_3 = B_2 w_1 - j_3/\sigma$$

$$\partial u/\partial x_1 = 0$$

It follows from these equations that w_1 can be measured by measuring $\partial u/\partial x_3$, if $j_3 = 0$ or is also measured. The case $j_3 = 0$ is realized when $\partial w_1/\partial x_2 = 0$, i.e., when the velocity gradient vanishes along the direction of the magnetic field. As a rule, j_3 cannot be measured directly because it appears in the equation that already contains the unknown velocity w_1. It follows that the conductive anomometer can be used in quantitative measurements only for flows with a uniform velocity distribution. However, this condition will not be satisfied in turbulent flows and the conductive anemometer yields only qualitative results, although these can serve as a basis for verifying different hypotheses about the effect of a magnetic field on turbulence.

The potential gradient is measured with an array of microelectrodes, arranged in a single probe which is inserted into the flow. The potential difference that appears between the electrodes is small. For a flow rate $w = 0.1$ m/s and $B = 1$ T, the potential gradient does not exceed 0.1 V/m, which gives $\Delta u \simeq 10^{-4}$ V when the electrode separation is 1 mm. The separation between the electrodes is therefore usually made to be a few millimeters, since it is only then that a useful signal can be extracted from electrochemical noise. However, the separation between the electrodes is not the only factor that determines the spatial resolution of the conductive anemometer. Thus, it is shown in [28] that the potential difference between the electrodes is significantly affected by the velocity distribution in the flow region, whose dimensions substantially exceed the separation between the electrodes.

In electromotive flows, the distribution of potential in the fluid is wholly determined by the high working current, and the small (a few tens of microvolts) signal generated by the conductive anemometer is practically indistinguishable from the background.

There are also doubts about the suitability of the hot–wire anemometer when the fluid is subjected to strong Joule heating, since small changes in the wire temperature, due to changes in the flow rate, are then masked by temperature fluctuations in the liquid metal at each point in the electromotive flow.

As noted in the Introduction, mechanical velocity transducers have many advantages in measurements under such complicated conditions. They are practically insensitive to the contamination of the sensitive element, and their transfer function is unaffected by thermal and electrical conduction in the working fluid. (We note that in strong magnetic fields, electrical conduction in the fluid may affect the calibration curve of the mechanical transducer). When high currents flow through the fluid, the only possible restrictions are those due to temperature and the temperature gradient at the point of measurement (see Section 11). Nevertheless, there is only one published attempt to use the mechanical transducer for quantitative measurements in a turbulent magnetohydrodynamic flow (in mercury) [8]. The authors of [8] used a strain transducer similar to that illustrated in Fig. 1c [33].

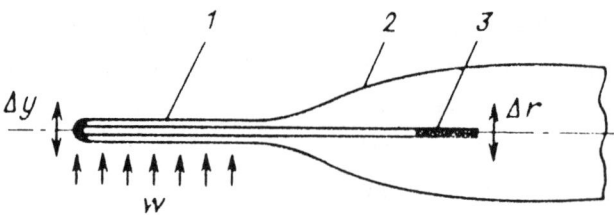

Fig.36 Principle of the sensitive element for an opaque liquid

Design of sensitive elements for opaque fluids. As already noted, the use of optical-fiber displacement devices has radically improved the characteristic of mechanical velocity transducers. However, the displacement transducer must be insulated from the opaque fluid. This can be done by exploiting the principle illustrated in Fig. 36. The sensitive element *1* of the velocity transducer is the end piece of a thin-walled glass capillary *2* (diameter about 0.5 mm) drawn out to a diameter of 0.3 – 0.5 mm. The glass fiber pointer *3*, blackened at the free end, is fused to the end of the sensitive element.

The displacement of the opaque end of the pointer by the flow is recorded by an optical-fiber displacement transducer similar to that shown in Fig. 25. The principles and the working characteristics of optical-fiber displacement transducers for opaque bodies are described in Section 3.

The fabrication technology available for producing the sensitive element with the pointer is relatively simple. The glass capillary is first placed in the device shown in Fig. 26. The conical shape of the sensitive element can be controlled by suitably choosing the heater current and the load. After the capillary has been pulled out, it is cut at the thick end and a glass fiber, 0.02 mm in diameter, is inserted into it (and will eventually act as the point). The best centering of the pointer in the sensitive element is produced by simple wedging of the fiber in the thin portion of the sensitive element. The segment of the sensitive element whose internal diameter is smaller than the diameter of the fiber is then snapped off (usually at the end of the pointer). The end of the sensitive element and the pointer are then fused by a microheater, and the pointer is blackened by inserting it into a capillary containing black ink. The last operation involves the mounting and adjustment of the displacement transducer relative to the pointer.

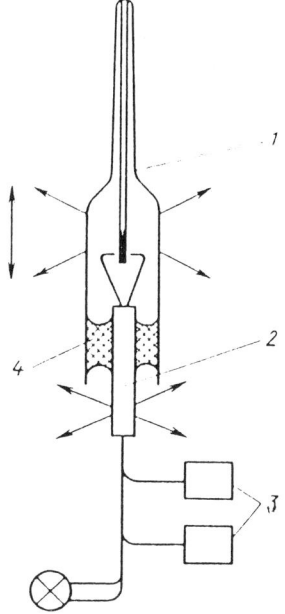

Fig.37 Adjustment of the velocity transducer

Figure 37 illustrates the adjustment of a two-component velocity transducer. The sensitive element 1 is held by a three-axis manipulator. The optical-fiber displacement transducer 2 is held by a two-axis micromanipulator which allows smooth displacement in the plane perpendicular to the axis of the sensitive element. A drop of an epoxy compound 4 is then introduced, and the manipulators are used to achieve the required mutual disposition of the components. The adjustment process is monitored by recording the signals from the photodetectors 3. The first step is to measure the signals from the two photodetectors with the pointer withdrawn from the measuring light beams. The manipulators are then used to reduce each signal by a factor of about 2, which ensures that the adjusted parameter is $\Delta r_0 \simeq 1$ and the displacement transducer operates on the linear portion of its characteristic (see Section 3). The system remains in the adjusted state until the compound solidifies. The assembled and adjusted velocity transducer is then inserted into the measuring probe and is calibrated in a flowing liquid metal (mercury).

Determination of the displacement of the pointer. When the sensitive element is inserted into a flow, its end is deflected by the amount Δy and its end cross section turns through an angle θ. The free end of the pointer is therefore displaced by the amount

$$\Delta r = L \sin \theta - \Delta y$$

Calculations performed for specific sensitive elements have shown that $\theta < 0.005$. Hence, $\sin \theta \simeq \theta_n$ and

$$\Delta r \simeq L \theta - \Delta y$$

where L is the length of the pointer.

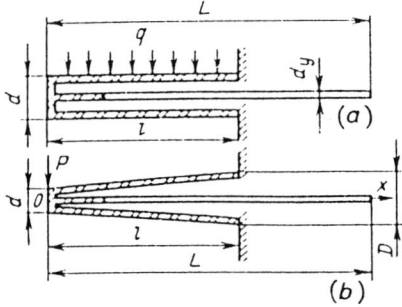

Fig.38 Models of sensitive element used in computations

The cylindrical model of the sensitive element (Fig. 38a) can be used to estimate the sensitivity of velocity transducers. We then have

$$\Delta y = \frac{ql^4}{8EI}$$

$$\theta = \frac{ql^3}{6EI}$$

$$\Delta r = \frac{ql^4}{24EI}(4\overline{L} - 3)$$

$$\overline{L} = \frac{L}{l}$$

At the end of the sensitive element

$$I = (\pi/64)\, d^4(1 - \beta^4)$$

where β is the ratio of the internal diameter of the sensitive element to its external diameter. Both the sensitivity and the resonance frequency increase with increasing β. In the first approximation, it may be considered that β remains constant when the thin-walled glass capillaries are pulled out. Good sensitive elements can be produced from glass capillaries for β > 0.9.

At this point, we need the formulas for the deformation of a conical cantilever beam loaded at the free end by a localized force P (Fig. 38b). Such formulas are not readily available in commonly used handbooks, so that we shall derive them from the differential equation for an elastic beam of variable cross section:

$$y''(x) = M(x)/[EI(x)]$$

The local bending moment and second moment of area of the cross section are then respectively given by

$$M(x) = Px$$

$$I(x) = (\pi/64)(1-\beta^4)d^4(x) = (\pi/64)(1-\beta^4)d^4(1+bx)^4$$

where $b = (D-d)/d$ and $\bar{x} = x/l$.

The cross section rotates through the angle

$$\theta(x) = y'(x) = \frac{APl^2}{d^4}\left[\int \frac{\bar{x}\,d\bar{x}}{(1+b\bar{x})^4} + C_1\right]$$

where $A = 64/\left[\pi(1-\beta^4)\right]$ and $\bar{x} = x/l$.

Integrating and applying the boundary condition $y'(\bar{x}) = 0$ at $\bar{x} = 1$ to obtain the integration constant C_1, we find that

$$\theta(\bar{x}) = \frac{APl^2}{d^4 b^2}\left[\frac{1+3b}{6(1+b)^3} - \frac{1+3b\bar{x}}{6(1+b\bar{x})^3}\right]$$

where for $x=0$

$$\theta(0) = \frac{APl^2}{D^4}\left[\frac{1}{3}\frac{D}{d} + \frac{1}{6}\left(\frac{D}{d}\right)^2\right] \tag{28}$$

The deflection is given by

$$y(\bar{x}) = l\int \theta(\bar{x})\,d\bar{x} + C_2$$

After the integration and the determination of the constant C_2 from the condition $y=0$ at $\bar{x} = 1$, we obtain

$$y(x) = \frac{APl^3}{b^2d^4}\left[\frac{1+3b}{6(1+b)^3} - \frac{1}{6b(1+b\bar{x})^2} + \right.$$

$$\left. + \frac{1}{6b(1+b)^2} + \frac{1}{2b(1+b\bar{x})} - \frac{1}{2b(1+b)}\right]$$

$$\Delta y = y(0) = \frac{APl^3}{3D^4}\frac{D}{d} \tag{29}$$

The deflection of the end of the pointer is

$$\Delta r = L\,\theta(0) - \Delta y$$

We must now find Δr for the case where the force P is applied at a distance x from the free end of the conical cantilever. At $\bar{x}$, we have

$$y(\bar{x}) = \frac{APl^3(1-\bar{x})^3}{3D^4}\frac{D}{d(\bar{x})}$$

$$\theta(\bar{x}) = \frac{APl^2}{D^4}\left[\frac{1}{3}\frac{D}{d(\bar{x})} + \frac{1}{6}\left(\frac{D}{d(\bar{x})}\right)^2\right]$$

At $\bar{x}=0$,

$$y(0) = y(\bar{x}) + x\,\theta(x), \quad \theta(0) = \theta(\bar{x})$$

$$\Delta r(\bar{x}) = L\theta(\bar{x}) - y(\bar{x}) - x\,\theta(\bar{x})$$

We now define the influence function by

$$\psi(\bar{x}) = \Delta r(\bar{x})/\Delta r(0)$$

and, using the above expression, we obtain

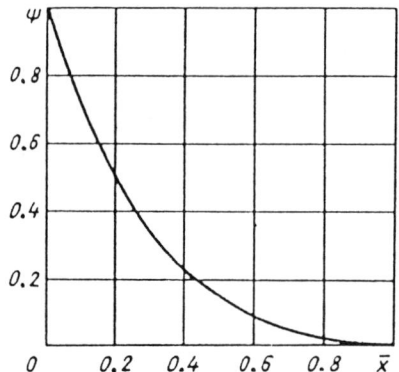

Fig.39 Graph of influence function for a finite beam ($b=1$, $\bar{L}=5$)

$$\psi(\bar{x}) = \left\{ \bar{L}\left(1-\bar{x}\right)^2 \left[\frac{1}{3} \frac{1+b}{1+b(\bar{x})} + \frac{1}{6}\left(\frac{1+b}{1+b\bar{x}}\right)^2 \right] \right.$$

$$- (1-\bar{x})^3 \frac{1+b}{3(1+b\bar{x})} - \bar{x}\left(1-\bar{x}\right)^2 \left[\frac{1}{3} \frac{1+b}{1+b\bar{x}} + \frac{1}{6} \times \right.$$

$$\left. \times \left(\frac{1+b}{1+b\bar{x}}\right)^2 \right] \right\} / \left\{ \bar{L}\left[\frac{1}{3}(1+b) + \frac{1}{6}(1+b)^2 - \frac{1}{3}(1+b) \right] \right\} \quad (30)$$

Figure 39 shows a graph of the function $\psi(\bar{x})$ for $b=1$ and $L=5$.

As noted in Section 9, the influence function characterizes the sensitivity distribution along the length of the sensitive element. As expected, the conical sensitive element has a still more nonuniform sensitivity distribution along the length than the cylindrical element (see Fig. 34). At $\bar{x}=0.6$, the influence function is $\psi(\bar{x}) \simeq 0.08$. The influence function is convenient in numerical calculations of Δr for an arbitary load distribution $q(\bar{x})$ along the length of the sensitive element. The analytic solution of this type of problem is very unwieldy, and is very rarely used in practical calculations.

Calibration in a liquid−metal flow. Different standardizing systems can be used to calibrate velocity transducers.

In a standardizing system of the first type [13], an axially symmetric nozzle is used to produce an immersed jet of mercury. The velocity is calculated from the known cross-sectional area of the nozzle and the rate of the flow of the fluid measured by determining the time taken to fill a tank of given volume. The clock is started and stopped by electrical contacts shorted by the mercury as the measuring tank fills up. This method can be used to determine the velocity to within about 0.3%. The transducer is placed in the zone of uniform velocity distribution of the nozzle, and the electrical signal is recorded as a function of velocity.

The standardizing system of the second type [21, 22] incorporates an arrangement for displacing the vertically mounted probe containing the velocity transducer in stationary mercury in a container of 20 × 50 × 640 mm. The container is placed between the poles of a magnet in the region in which the magnetic field is uniform. The transducer and the primary electronics are mounted on a platform that can be displaced by a dc motor. Special measures are taken to prevent vibration due to the working of the motor and the uniform displacement of the transducer in mercury is monitored. The sensitive element is at a distance of 6 − 7 mm from the free surface of the mercury. The transducer traverses a distance of 500 mm in each measurement. The velocity is measured over a 200 mm segment in which the velocity of the transducer relative to the liquid mercury remains constant to within 0.5%. The velocity is determined by dividing the length of the segment by the time measured with a frequency meter that is started and stopped by pulses generated and recorded at the ends of the 200-mm segment.

An integrating voltmeter (see Section 19), based on a voltage-to-frequency converter and calibrated against a dc voltmeter, is used to determine the average signal from the transducer. The pulse repetition frequency generated by this device is directly proportional to the transducer signal. Consequently, the ratio of the number of pulses recorded during the measurements to duration of the measurements is proportional to the signal generated by the velocity transducer, averaged over the interval. The use of this integrator significantly reduces the parasitic signal due to noise in the electronics, the electromagnetic pick-up, and the vibrations of the system. The pulse counter is started and stopped by the same electronics that controls the clock.

A system incorporating a velocity transducer floating on stationary mercury can be used to calibrate the transducer in a transverse magnetic field. However, calibration across the aperture of a nozzle cannot be carried out in a magnetic field perpendicular to the velocity because this upsets the axial symmetry of the jet of mercury and introduces errors into the velocity determined from the rate of flow. Velocity transducers can be calibrated in liquid metals using the system proposed in [36] in which the motion of the velocity transducer relative to a stationary liquid is determined by a massive pendulum carrying the transducer. The velocity at each instant of time is determined from the measured amplitude and the known period of the pendulum. The velocity of the transducer relative to the stationary liquid varies between zero and its maximum value in a time equal to one quarter of the period of the pendulum. The maximum velocity is preset by the initial deflection of the pendulum from its equilibrium position. The measurements are thus performed dynamically, which means that the recording equipment must have sufficient time resolution.

The velocity transducers incorporating optical-fiber displacement transducers (see Fig. 25) are illustrated together with typical calibration curves. The two sets of points in Fig. 40 corresponds to Figs. 25a and b, respectively. It is clear from Fig. 40 that both calibration curves can be described by (23) with $n \simeq 1.6$. Since the displacement transducer in which the lightguides are cut at 45°, and are metalized, has a linear relationship between the output signal and the displacement of the pointer, we may conclude that the displacement transducer with curved cylindrical lightguides must also have a linear characteristic. This is indicated by the

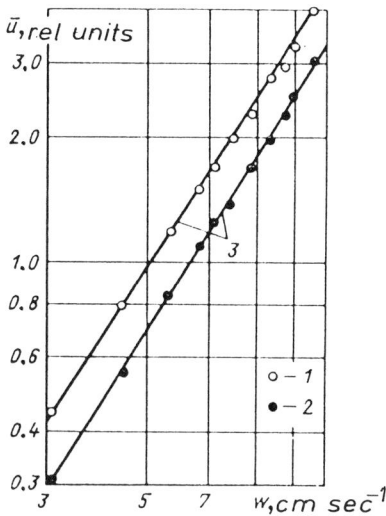

Fig.40 Calibration curves of a velocity transducer for flowing mercury

good agreement between experimental data and calculations based on (14), (15), and (17) (curves 3).

Influence of the magnetic field. The effect of magnetic field on the calibration of optical-fiber velocity transducers was investigated in [21] and [36]. In [21], the transducer was towed at a constant rate across a pool of mercury placed in the gap of a magnet. The magnetic induction B was perpendicular to the velocity of the transducer and to the axis of the sensitive element (transverse field). The measurements were first performed in zero field. The magnet was then turned on and the measurements were repeated.

Figure 41 shows the results obtained for velocities between 2.9 and 11 cm/s in a magnetic field of 0.63 T. The diameter and length of the sensitive element of the transducer were $d \simeq 0.05$ mm and $l \simeq 1$ mm, respectively. It is clear that the possible influence of the magnetic field does not exceed the experimental uncertainty ($\pm 4\%$).

It is well known [58] that the effect of the magnetic field on the flow of

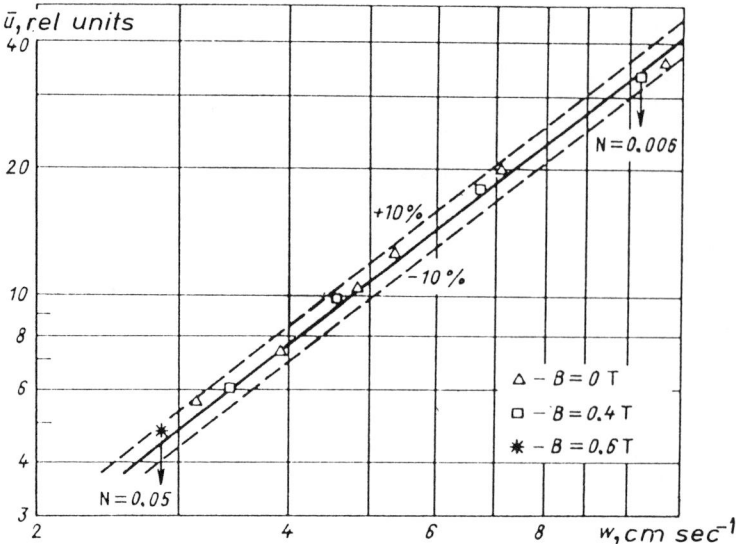

Fig.41 Calibration curve of a velocity transducer for mercury in a magnetic field

an electrically conducting liquid past a given body can be characterized by the dimensionless number

$$N = B^2 \sigma d / w \rho_l$$

where σ and ρ_l are, respectively, the electrical conductivity and the density of the liquid and d is the diameter of the sensitive element.

It is found that $0 < N < 0.05$ for the above values of w and B.

Similar measurements were performed in the range $0 < N < 0.02$ using a pendulum system. The magnetic field could be either parallel or perpendicular to the velocity. The magnetic field was found to have no perceptible influence on the calibration curves.

The results obtained in [21] and [36] are in agreement with the calculated drag coefficients of a circular cylinder placed transversely to a flow in transverse and longitudinal magnetic fields for $Re_d = 40$ and $N < 1$

[35]. In the case of the transverse field, calculations show that

$$C_d(B)/C_d(0) = 1 + 1.08N$$

The influence of the longitudinal magnetic field is weaker:

$$C_d(B)/C_d(0) = 1 + 0.25N$$

The calculations were performed for cylinders of infinite length. Measurements were then made of the influence of the relative length $\bar{l} = l/d$ of the cylinder on the drag coefficient in a magnetic field [58]. It was found that the coefficient of N in the above formulas for $\bar{l} \simeq 20$ had to be reduced by a factor of about 2. This means that the increase in the transducer signal should not exceed about 3% for the maximum value $N = 0.05$ used in these experiments.

Since the effect of the magnetic field on the calibration curves of optical-fiber transducers is slight, this simplifies the calibration procedures as well as measurements in turbulent magnetohydrodynamic flows.

Measurement of the two velocity components. As noted previously (see Section 9), the velocity transducer using two crossed optical-fiber displacement transducers can be used to measure the two components of the velocity that lie in the plane perpendicular to the axis of the sensitive element. The necessary condition for such measurements is that the two channels of the displacement transducer must be independent. This in turn means that, first, each of the displacement transducers must not react to the displacement of the pointer along its optical axis and, second, the two channels do not illuminate one another. These conditions can be verified experimentally as follows.

The measuring probe with the two-component velocity transducer is held in an apparatus that enables it to be rotated around the axis of the sensitive element and the angle of rotation ϕ can be measured. Next, the signal u_i produced in each of the transducer channels is measured at

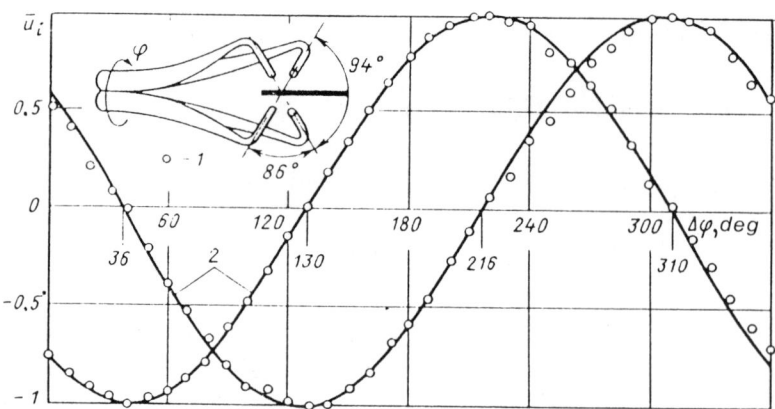

Fig.42 The output of a velocity transducer for mercury as a function of the angle of rotation

constant flow rate as a function of the difference $\Delta\phi = \phi - \phi_0$ where ϕ_0 is some initial value of the angle of rotation, which can be arbitrary. The measured values of u_i are then used to determine the dimensionless normalized values given by

$$\bar{u}_i = (u_i - u_{i\,0})/(u_{i\,max} - u_{i0})$$

where u_{i0} is the signal in the i-th channel for $w = 0$ and $u_{i\,max}$ is the maximum value of u_i for $0 < \Delta\phi < 360°$.

Figure 42 shows graphs of $\bar{u}_i(\Delta\phi)$ for a particular velocity transducer and compares measurements (open circles) with calculations (curves marked *2*) based on the formulas

$$\bar{u}_i = \sin(\Delta\phi - \delta_i) \tag{31}$$

The initial phases δ_i are determined for each channel from the positions of the zeros of the functions $u_i(\Delta\phi)$. It is clear from the Figure that, in this case, $\delta_1 = 130°$ and $\delta_2 = 216°$. It is also clear that the difference $\delta_2 - \delta_1$ is equal to the angle between the optical axes of the crossed displacement transducers. For this particular velocity transducer, this

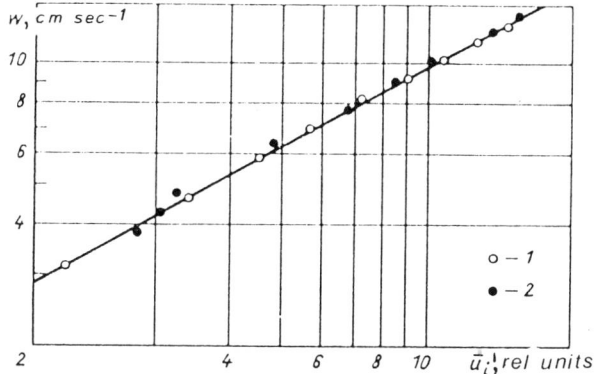

Fig. 43 Calibration of the two channels of the velocity transducer in flowing mercury

angle differed from the right angle by about 4°. The fact that the calculated and measured values of the functions $u_i(\Delta\phi)$ were found to agree confirms that the two cross displacement transducers were in fact independent and that there was no mutual illumination of the channels.

The two-component velocity transducer tested in this way is then calibrated for the two velocity components. The calibration is carried out in a standardizing system in which the transducer is pulled across a stationary pool of a mercury. The angle $\psi_{1max} = \delta_1 + 90°$ is first preset by a rotating device and the function $\overline{u_1}(w)$ is recorded (points 1 in Fig. 43). The angle $\psi_{2max} = \delta_2 + 90°$ is then set, and the velocity signal $\overline{u_2}(w)$ is recorded for the second channel (points 2, Fig. 43). It is clear from the figure that the functions $u_i(w)$ can be represented by (23) with $n \simeq 1.77$ in both channels. The fact that the values of n are roughly the same for the transducer channels shows that the characteristics of the displacement transducers are linear. Consequently, the two orthogonal velocity components w_1 and w_2 can be determined from u_1 and u_2 by the algorithm described in Section 9 [formulas (23) and (24)]:

$$\overline{u}_1 = c_1 w_1^{1.77}$$

$$\overline{u}_2 = c_2 w_1^{0.77} w_2 \qquad (32)$$

During the calibration process, the two perpendicular velocity components w_1 and w_2 are defined by the angle $\Delta\phi = \phi - \delta_1 - 90°$, so that $w_1 = |w|\cos\Delta\phi$ and $w_2 = |w|\sin\Delta\phi$.

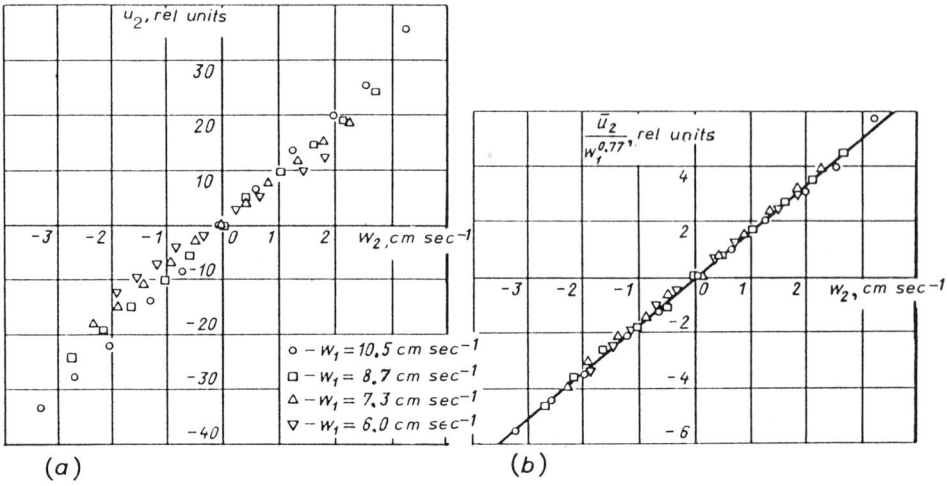

Fig.44 The component $\overline{u_2}$ as a function of w_2 for different w_1 (a) and the results based on (32) (b)

The first step is to verify that the second channel is orthogonal to the velocity w_1. The angle $\Delta\phi$ is measured from the point at which $u_2 - u_{20} = 0$ for all w_1. Figure 44a shows the measured values of $\overline{u_2}$ for w_1 between 6 and 10.5 cm/s and w_2 between -3.2 and $+3.2$ cm/s. It is clear that the function $\overline{u_2}(w_2)$ is not single valued, and that the experimental points are spread over the values of w_1, as expected from (32). The experimental results can be described by

$$\overline{u_2}/w_1^{0.77} = f(w_2)$$

which means that a single calibration curve can be obtained for all values of w_1 (Fig. 44b), which shows that (32) is valid in this case and can be used to implement a simple algorithm for processing phototransducer data u_1 and u_2 with the view to an independent determination of the two orthogonal velocity components w_1 and w_2 (see Section 19).

When the sensitivities of the two displacement transducers are equal ($c_1 = c_2 = c$), the velocity transducer is more sensitive to the "longitudinal" component w_1. In actual fact,

$$\partial \overline{u_1}/\partial w_1 = cn\, w_1^{n-1}$$

$$\partial \overline{u_2}/\partial w_2 = c w_1^{n-1}$$

Since $n = 1.6-1.85$ for all the transducers that have been investigated, measurements in turbulent flows are best carried out so that the velocity transducer is oriented in such a way that the velocity component perpendicular to the mean flow velocity is measured in the most sensitive transducer channel. This minimizes the experimental uncertainty in the measured transverse turbulent velocity fluctuations.

The above spread in the values of the exponent n is due above all to the diameter and configuration of the tip of the sensitive element. Other things being equal, the minimum value of n corresponds to the case where the sensitive element gradually comes to a point as shown in Fig. 36. Occasionally, when the pointer is fused to the tip of the sensitive element, its spherical shape becomes flattened and, if its diameter exceeds by a factor of two the initial diameter of the tip, the value of n rises to 1.85. An increase in n is often desirable, for example, in measurements of the so-called Reynolds *frictional stress* $<w_1'\, w_2'>$, which is equal to the time average of the product of the longitudinal and transverse components of turbulent velocity fluctuations. It is then found that

$$<w_1'\, w_2'> = <w_1 w_2> = c<\overline{u}_1^{2/n-1}\, \overline{u}_2>$$

When $n \to 2$, we have $<w_1 w_2> \to c<\overline{u}_2>$, i.e., the Reynolds stresses can be determined from the time average of the signal $\overline{u}_2$.

Effect of nonorthogonality of optical axes. Practical velocity transducers suffer from assembly errors whereby the angle between the optical axes of the crossed displacement transducers differs from the right angle by some small amount β. The velocity transducer must then be oriented so that the experimental uncertainty in the two velocity components is a minimum. Two possible orientations are shown in Fig. 45.

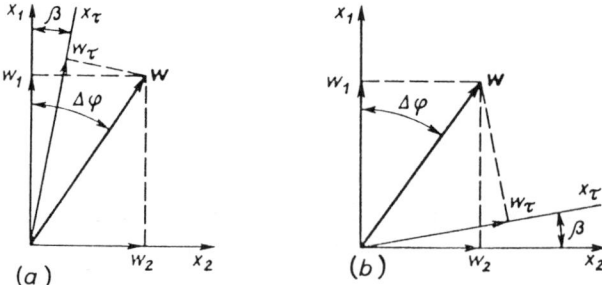

Fig.45 Different orientations of a nonorthogonal velocity transducer relative to the flow

In the first case (Fig. 45a), the optical axis of one of the channels lies along the x_2 axis which is perpendicular to the velocity component $<w_1>$. The second optical axis x_τ is at an angle β to w_1. This arrangement measures the components w_τ and w_2 and, when w_2 is measured "accurately", and w_τ differs from w_1, we have

$$w_\tau/w_1 = \cos\beta \pm |w_2/w_1| \sin\beta$$

When measurements are performed in a turbulent flow and the average velocity points along the x_1 axis we have $w_2/w_1 < 1$, and both positive and negative values of w_2 are equally likely. The error introduced by replacing w_1 with w_τ is a maximum for negative w_2 (minus sign on the right-hand side of the above relation). Figure 46a shows the results of calculations for this case for different values of the ratio w_2/w_1. The uncertainty in the measured w_1 due to the nonorthogonality of the transducer is relatively large: it amounts to 4% for β = 6° and $w_2/w_1 = 0.3$, i.e., it is the same as the error introduced by replacing $|w|$ with w_1.

If the nonorthogonal transducer is oriented so that one of its optical

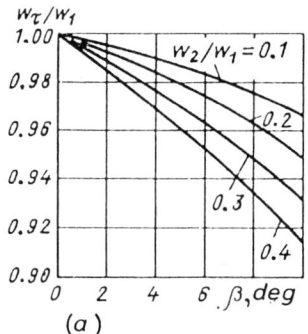 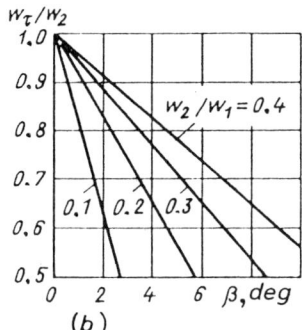

Fig.46 Uncertainties in measured w_1 (a) and w_2 (b) as functions of the angle β and the ratio w_2/w_1

axes lies along the x_1 axis (see Fig. 45b), the error in the measured second velocity component w_2 under the same conditions is relatively large:

$$w_\tau/w_2 = \cos \beta \pm |w_1/w_2| \sin \beta$$

The negative sign in this expression corresponds to the maximum error in the measured w_2. Since $w_1/w_2 > 1$, the error in the measured w_2 is greater by about an order of magnitude than the error in the measured w_1 in the last case. It is clear from Fig. 46b, which shows the results based on the last equation, that quantitative measurements are generally impossible in this case even for $\beta = 2°$. The velocity transducer must therefore be oriented in the turbulent flow so that one of its axes is exactly parallel to the x_2 axis.

In many cases, e.g., in measurements in the boundary layer in the presence of secondary flow, the average value $<w_2>$ can be different from zero. When the fluctuation velocity component w_2 is the only one to be measured in this case, the transducer must be oriented so that

$$\int_0^T w_2(\tau)\, d\tau = 0$$

i.e., the constant component $<w_2>$ should be compensated by the component of $<w_1>$ along the x_τ direction. Provision for this must be made in the measuring equipment. The integration time T must be long enough (significantly greater than the period of the low-frequency energy-bearing component fluctuations $\bar{w_2}$).

There is also another way: the constant component can be excluded by the use of suitable filters in the electronic circuitry. However, this unavoidably produces a cut off in the low-frequency components, which may lead to some loss of information about the energy-bearing low-frequency part of the spectrum in the case of measurements in liquid metals.

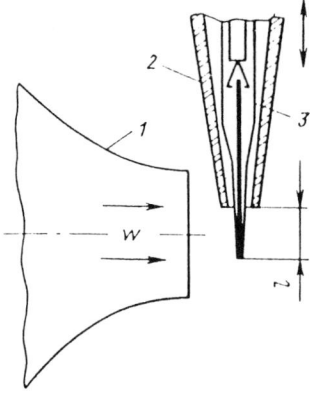

Fig.47 Arrangement used to measure L_{eff}

Spatial resolution. In the case of measurements of turbulent fluctuations in hydrodynamic variables, the dimensions of the sensitive element of the transducer must be significantly smaller than the spatial scale of the fluctuations. Unless this is so, the transducer will record the average over small-scale fluctuations, which may result in significant errors in the experimental data, or even qualitatively incorrect conclusions drawn from them [52].

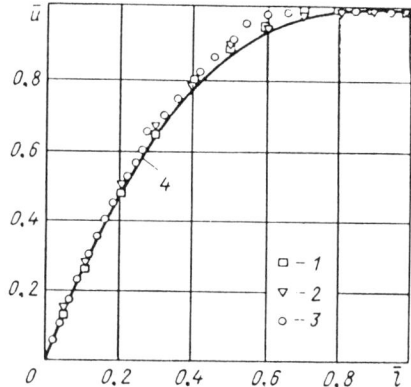

Fig.48 The function $\bar{u}\,(\bar{l})$ for two specimens of velocity transducers

Since the spatial scale of fluctuations is related to their frequency [52], the dimensions of the transducer usually determine the upper limit of the frequency range within which measurements can be made.

Transducer dimensions are also significant in relation to measurements on nonuniform flows that are characterized by high velocity gradients. The so-called reference error is important in this case even when time-averaged quantities are measured. The point is that the coordinate of a point to which a particular measurement is referred can be determined only to within an uncertainty of the order of the dimensions of the sensitive element. When the position coordinate (e.g., distance from the wall) is comparable with the dimensions of the sensitive element of the velocity transducer, qualitative measurements of the velocity gradient become impossible.

The spatial resolution of a transducer must therefore be as high as possible, i.e., the volume $v = (\pi d^{2}/4) l_{eff}$ in which the transducer determines the required variable must be as small as possible (d is the diameter and l_{eff} is the effective length of the sensitive element).

In the case of the velocity transducers for opaque media considered here, the value of l_{eff} is undetermined because the diameter of the sensitive element varies smoothly from its minimum value at the tip of the element to the diameter of the holder. The value l_{eff} must therefore be found experimentally.

The length l_{eff} is measured in the uniform flow of air across the aperture of the axially symmetric nozzle *1* (Fig. 47). The velocity transducer *3* is shielded from the flow by a hollow glass cone *2* whose exit aperture has a diameter of 0.5 mm. The transducer is displaced along the cone axis, and the displacement can be measured to within ±0.001 mm with a traveling microscope. The signal from one of the transducer channels is recorded as a function of the length l of the sensitive element protruding into the flow. The flow rate is varied between 8 and 40 m/s, and its uniformity over the cross section of the jet is verified with a Pitot tube. There are no velocity gradients in the zone in which measurements are made.

Two specimen velocity transducers with slightly different dimensions of the sensitive element were examined. In transducer No.1, the diameter of the tip of the sensitive element was approximately 0.04 mm; it increased by a factor of 2 over the distance $l_2 \simeq 1$ mm from the tip. Transducer number 2 had $l_2 \simeq 1.2$ mm. Measurement showed that the signal did not increase with increasing l for $l > l_2$. This means that the sensitive element could be looked upon as a conical cantilever of length l_2, clamped rigidly at the thick end.

Figure 48 shows the experimental results obtained for the two transducers [*1* and *2* refer to transducer No.1 and $w = 5$ and 9 m/s, respectively; *3* refers to transducer No. 2 and $w = 40$ m/s; 4 is calculated from (33)]. The dimensionless signal strength was defined by $\bar{u} = u/u_{max}$ where u_{max} is the maximum value of u for $l > l_2$. The scale for l was taken to be l_2 ($\bar{l} = l/l_2$). The experimental points obtained for the two transducers lie on a single curve. The flow rate was varied in a wide range, but was found to have little effect on these results.

The function $\overline{u}(\overline{l})$ was calculated from (30) using the influence function ψ. The constant b in (30) was equal to unity in this case. It is clear that

$$\overline{u}(\overline{l}) = \int \psi(\overline{x})\, q(\overline{x})\, d\overline{x} \Big/ \int \psi(\overline{x})\, q(\overline{x})\, d\overline{x} \qquad (33)$$

where the first integral is evaluated between 0 and $\overline{l}$, and the second between 0 and 1.

Since $q \sim d$ for a cylinder placed transversely in the flow, it is natural to assume in this case that

$$q(\overline{x}) \sim d(\overline{x})$$

or

$$q(\overline{x}) = q(0)\,(1 + \overline{x})$$

where $\overline{x} = x/l_2$. The function $\overline{u}(\overline{l})$ calculated from (33) is shown in Fig. 48 (curve 4). It agrees with the experimental data.

The data shown in Fig. 48 can be used to estimate l_{eff}. Taking $\overline{l}_{eff}$ as being the value of $\overline{l}$ for which $\overline{u} = 0.95$, we find that $\overline{l}_{eff} = 0.55 - 0.6$, so that for this particular transducer $l_{eff} \simeq 0.6$ mm. The averaging volume within which the two velocity components are determined does not then exceed 10^{-3} mm^3. The optical-fiber transducer is thus seen to have very high spatial resolution which, so far, is matched only by the laser Doppler anemometer [17] in the case of measurements in transparent liquids. In the case of liquid metals, the spatial resolution of the hot-wire anemometer with crossed wires is worse by several orders of magnitude (and by an even greater figure in the case of the conductive anemometer).

We note that the above estimate of v must be looked upon as an upper limit because the sensitivity is distributed very uniformly over the length of the conical sensitive element in [see the graph of $\psi(\overline{x})$ in Fig.39]. The good agreement between calculated and experimental data shows that l_{eff} can be estimated by simple numerical procedures.

11. OPTICAL–FIBER VELOCITY TRANSDUCERS IN FLOWS WITH VELOCITY AND TEMPERATURE GRADIENTS

Optical-fiber velocity transducers have high spatial resolution which means that the two velocity components can be measured in the immediate vicinity of the walls of a tube, where the hot–wire anemometer [46] or even the laser Doppler anemometer encounter considerable difficulties that become even greater when there is heat transfer between the fluid and the wall. The velocity falls rapidly as the wall is approached and there is a simultaneous increase in the velocity (and temperature) gradients that influence the transducer operation [20]. We shall confine our attention to the so-called viscous sublayer $0 < x_2^+ < 5$ in which the velocity distribution along the normal to the wall is linear:

$$w_1(x_2) = u^+ x_2^+$$

where $u^+ = (\xi/8)^{1/2}\overline{w_1}$ is the dynamic velocity, $x_2^+ = x_2 u^+/\nu_l$, x_2 is the distance between the wall and the point at which the velocity is measured, $\overline{w_1}$ is the velocity averaged over the cross section of the tube, $\xi = 0.316/\overline{Re}^{0.25}$ is the coefficient of hydraulic friction [30], ν_l is the kinematic viscosity of the fluid, $\overline{Re} = \overline{w_1} d_e/\nu_l$, and d_e is the equivalent hydraulic diameter of the tube.

Consequently,

$$x_2^+ = 0.20\, \overline{Re}^{0.875}\, x_2/d_e \tag{34}$$

$$w_1(x_2) = 0.20\, \overline{Re}^{0.875}\, x_2^+ \nu_l/d_e \tag{35}$$

$$u^+ = 0.20 \,\overline{\text{Re}}^{-0.875} \nu_l/d_e \tag{36}$$

It follows from (35) that for $\overline{\text{Re}} = 10^4$ and $d_e = 50$ mm, we have $w_1(x_2) \simeq 0.2 x_2^+$ mm in air. To measure the velocity for, say, $x_2^+ = 0.1$, we have to be able to measure values of the order 1 cm/s. Optical-fiber velocity transducers such as those of Figs. 28 and 30 have no lower sensitivity limits and can therefore be used to perform such measurements with the sensitive element (of suitable length) placed parallel to the wall surface.

Influence of the velocity gradient. A cylindrical sensitive element placed in a flow with a velocity gradient experiences a drag and also a lift in the direction in which the velocity is increasing. The influence of the latter on the measurements performed with the hot-wire anemometer in the vicinity of the wall was investigated in [53]. The lift tends to bend the wire, which increases the effective distance between the wire and the wall. The lift was calculated in [53], and the experimental data obtained for the turbulent boundary layer were corrected accordingly. This removed the discrepancy between theory and experiment. We shall use this method to estimate the uncertainties in the measured velocity that are due to the velocity gradient.

According to [53], the lift per unit length of a cylindrical sensitive element is

$$q_2 = 0.5 \, \rho_l \, d \, w_1^2(x_2) \, G \tag{37}$$

where G is the dimensionless velocity gradient.

In the viscous sublayer, $w_1 \sim x_2$ and $G = d/x_2$. The drag is given by

$$q_1 = 0.5 \, C_d(\text{Re}_d) \, \rho_l \, w_1^2(x_2) \, d \tag{38}$$

where $C_d(\text{Re}_d)$ must be calculated from (11) since $\text{Re}_d \ll 1$ in the viscous sublayer.

The lift force can be characterized by the ratio q_2/q_1. Using (11) and (34) – (38) we obtain

$$q_2/q_1 \simeq 0.005\, \overline{\mathrm{Re}}^{-1.75} (d/d_e)^2 \ln(37.5 d_e/\, x_2^+\, \overline{\mathrm{Re}}^{-0.75}\, d\,) \qquad (39)$$

As example, consider the flow of air ($\rho_l \simeq 1.2$ kg/m^3, $\nu_l = 1.5 \times 10^{-5}$ m^2/s) in a tube with $d_e = 5 \times 10^{-2}$ m and $\overline{\mathrm{Re}} = 10^4$. Assuming that $d = 0.005$ mm and that the minimum distance between the wall of the tube and the axis of the sensitive element is $x_{2min} = 0.005$ mm, we find from (34) – (36) that $x_{2min}^+ = 0.06$. An optical-fiber velocity transducer similar to that shown in Fig. 30 can therefore be used to measure (at least in principle) the two velocity components much closer to the wall than the hot-wire anemometer (for which $x_{2min} \simeq 1$ mm) or even the laser Doppler anemometer which has been used under similar conditions only for $x_2^+ > 1$. It follows from (39) that the value of q_2/q_1 corresponding to $x_2^+ = 0.06$ does not exceed 0.005 in this case, i.e., the velocity gradient has no perceptible influence on the operation of the transducer.

The above method of estimating the influence of the velocity gradient can be used in a wide range of working liquids and experimental conditions.

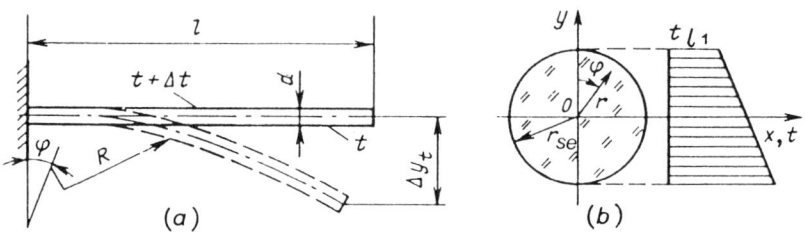

Fig.49 Deflection of a nonisothermal sensitive element

Influence of temperature gradient. A gradient in the temperature t_l of the liquid, and the associated temperature variation along the sensitive element, give rise to a lower effective temperature. The cylindrical sensitive element (Fig. 49a) of length l and diameter d is heated nonuniformly: the temperature on the upper generator is higher by the amount Δt. Because the uppermost generator becomes much longer, the sensitive element is deformed into a circular arc of radius R. The end of the sensitive element is thus displaced by the amount $\Delta y_t = R(1 - \cos \phi) = 2R \sin^2 \frac{1}{2} \phi \simeq \frac{1}{2} R\phi^2$ where

$$l(1 + \gamma \Delta t) = (R + d) \phi$$

$$l = R\phi$$

and γ is the thermal expansion coefficient of the material of the sensitive element.

Solving the last three equations for ϕ, R, and Δy_t, and rearranging, we obtain

$$\Delta y_t = \gamma(\Delta t) l^2 / 2d$$

The determination of the influence of the temperature gradient $\partial t_l / \partial x_2$ on the thermal displacement of the sensitive element is thus reduced to the determination of the dependence of Δt on $\partial t_l / \partial x_2$. In a turbulent flow near the wall, the temperature gradient is

$$\partial t_l / \partial x_2 = q_c / \left[\lambda_l \left(1 + \Pr \epsilon_\tau / \Pr_t \nu_l \right) \right]$$

where q_c is the thermal flux density at the wall, λ_l is the thermal conductivity of the liquid, Pr is the Prandtl number of the liquid, and the turbulent Prandtl number $\Pr_t$ can be set equal to unity. The dimensionless turbulent momentum-transfer coefficient can be calculated from the Reichardt formula for the viscous sublayer:

$$\epsilon_\tau / \nu_l = 0.4 \left[x_2^+ - 11 \tanh \left(x_2^+ / 11 \right) \right]$$

The temperature distribution of the liquid over the perimeter of the sensitive element (Fig. 49b) is

$$t_l(\phi) = t_{l1} + \frac{\partial t_l}{\partial x_2} r_0 (1 - \cos \phi)$$

where t_{l1} is the minimum temperature (for $\phi = 0$).

The temperature field over the cross section of the sensitive element is described by the equation of thermal conduction

$$\frac{\partial^2 t}{\partial r^2} + \frac{1}{r} \frac{\partial t}{\partial r} + \frac{1}{r^2} \frac{\partial^2 t}{\partial \phi^2} = 0$$

subject to the boundary condition

$$-\lambda_{se} \left(\frac{\partial t}{\partial r} \right)_{r_0, \phi} = \alpha \left[t(r_0, \phi) - t_{l1} - \frac{\partial t_l}{\partial x_2} r_0 (1 - \cos \phi) \right]$$

where λ_{se} is the thermal conductivity of the material of the sensitive element (glass), α is the heat transfer coefficient, and $r_0 = d/2$.

Using the standard method of solving this type of boundary-value problem [9], and assuming that α is constant over the perimeter, we obtain

$$t(r, \phi) = t_{l1} + \frac{\partial t_l}{\partial x_2} \frac{d}{2} \left(1 - \frac{2r}{d} \frac{\text{Bi}}{1 + \text{Bi}} \cos \phi \right)$$

where $\text{Bi} = \alpha d / 2\lambda_{se}$ is the Biot number.

The maximum temperature drop over the cross section of the sensitive element is

$$\Delta t_{max} = t(r_0, \pi) - t(r_0, 0) = \frac{\partial t_l}{\partial x_2} d \frac{\text{Bi}}{1 + \text{Bi}}$$

where Bi= $\mathrm{Nu}_l \lambda_l/(2\lambda_{se})$, $\mathrm{Nu}_l = 0.42\, Pr_l^{0.2} + 0.57\, \mathrm{Pr}_l^{0.43}\, \mathrm{Re}_l^{0.5}$ is the Nusselt number determined from the formula for the heat transfer from cylinders placed transversely in a flow [30], $\mathrm{Re}_l = w(x_2)d/\nu_l$, and $\mathrm{Nu}_l = \alpha d/\lambda_l$.

The above formulas can be used to estimate the influence of a temperature gradient on the thermal deflection of the sensitive element. For the above examples of air flowing in a tube, in which $q_c = 10^4$ W/m^2, $x_2^+ = 0.06$, $\lambda_{se} = 0.7$ W/m.K, $\gamma = 5 \times 10^{-6}\, K^{-1}$, $d = 5 \times 10^{-6}$ m, and $l = 2 \times 10^{-3}$ m, we obtain $\Delta y_l/\Delta r_1 \simeq 0.027$, which is acceptable for optical–fiber velocity transducers in transparent fluids when measurements are made near the heated tube wall (Δr_1 is the deflection along the x_1 axis under the influence of the velocity).

Analysis of the above relationships shows that, in the first approximation, $\Delta y_l \simeq d$. Consequently, the uncertainty in the measured velocity in flows with a temperature gradient lies within acceptable limits, but only in the case of the velocity transducers discussed in Section 9 for transparent liquids in which the illuminating lightguide is used as the sensitive element. For opaque fluids, the characteristic value of d must be taken to be the diameter of the capillary 2 (see Fig. 36), which is ~ 0.5 mm. The uncertainty obtained above must then be increased by a factor of at least 100. Consequently, velocity transducers for opaque media are not very suitable for measurements in flows with a temperature gradient.

CHAPTER
FOUR

DYNAMIC CHARACTERISTICS OF OPTICAL–FIBER VELOCITY AND PRESSURE TRANSDUCERS

12. MATHEMATICAL DESCRIPTION OF THE OSCILLATIONS OF SENSITIVE ELEMENTS

The sensitive elements of velocity and pressure transducers (cantilevers and membranes, respectively) that are used to measure time–dependent velocities and pressures take part in a complicated oscillatory process whose properties are determined by external factors and by the properties of the sensitive elements themselves. The mathematical description of this process can be obtained, as usual, by considering all the forces acting on an element of the system. Modern theories of physical processes accompanied by oscillatory phenomena under real conditions involve the consideration of elastic forces P_e, inertial forces P_i, damping forces due to the external medium P_d, and damping forces P_f due to internal friction in the material. In the case of the sensitive element of a velocity transducer for transparent media (rods of constant cross section), these forces can be determined as follows.

The elastic force is given by

$$P_e = EI\, \partial^4 y/\partial x^4 \qquad (40)$$

where E is Young's modulus, I is the second moment of area of the cross section of the rod, x is the distance along the axis of the rod, and y is the deflection of the axis from its position of equilibrium.

The inertial force has two components:

$$P_{i1} = \rho F\, \partial^2 y/\partial \tau^2$$
$$P_{i2} = -\rho I\, \partial^4 y/\partial \tau^2 \partial x^2 \qquad (41)$$

The first of these requires no explanation and the second is due to the rotation of the cross section of the rod and can be neglected for long rods ($l/d > 10$).

Damping forces due to the resistance of the external medium can be determined from (15), written in the form

$$P_d = 0.5\, C_d\, \rho_l (dy/d\tau)^2\, d$$

However, the resulting differential equation for the oscillations is then nonlinear. It is common to use the linear approximation in which

$$P_d = k\, dy/d\tau = k\, \dot{y} \qquad (42)$$

Internal friction in the material of the rod is the least known. However, we do know that the most important phenomena responsible for the damping of the oscillations are plastic deformations, intercrystalline atomic diffusion (in the case of very small deformations), magnetoelastic hysteresis (in ferromagnets), and thermoelastic effects [57].

The essence of the thermoelastic effect is as follows. As the rod oscillates, its individual filaments are alternately stretched and compressed. During this process, the temperature of the compressed filaments rises and that of the extended filaments falls, so that there is an additional bending moment due to the nonuniform thermal expansion (see Section 11). Because of thermal conduction, this phenomenon is irreversible, i.e., the energy used to heat the compressed filaments is not fully recovered at a subsequent phase of the oscillations. This can give rise to peculiar resonance effects when the period of the oscillations and the thermal-wave propagation time at right angles to the rod axis are equal.

Since the internal friction effects have not been adequately investigated and they produce relatively small damping, it is common to employ semi-empirical formulas for P_f. If the differential equation for the oscillations then turns out to be linear, so much the better. Hooke's law (the relationship between stress and strain) is often taken in the generalized form [57]

$$\sigma = E[(\mu_0 + \mu_1 \operatorname{sign} \epsilon_0 + \mu_2 \epsilon_0^2)\, \partial \epsilon/\partial \tau + \xi\, \partial^2 y/\partial x^2]$$

where σ is the stress, ϵ the strain, ϵ_0 the amplitude of ϵ, and ξ the distance between a given filament and the neutral axis.

We then find that sign $\epsilon_0 = -1$ for $\epsilon_0 < 0$ and sign $\epsilon_0 = +1$ for $\epsilon_0 > 0$. If we suppose that internal damping is independent of the oscillation amplitude ($\mu_1 = \mu_2 = 0$), then

$$P_f = \mu_0\, EI\, \partial^5 t/\partial \tau \partial x^4 \tag{43}$$

Using (40) – (43), we obtain a linear differential equation for the forced oscillations of a rod of constant cross section:

$$EI\, \frac{\partial^4 y}{\partial x^4} + pF\, \frac{\partial^2 y}{\partial \tau^2} + \mu_0\, EI\, \frac{\partial^5 y}{\partial \tau \partial x^4} + k\, \frac{\partial y}{\partial \tau} = P(x, \tau) \tag{44}$$

where $P(x, \tau)$ is the external force.

The boundary conditions are specified at the ends of the rods and depend on the way they are clamped: at the free end $y'' = y''' = 0$ and at the rigidly clamped end $y = y' = 0$.

An analytic solution can be obtained for (44) [57], but it is so unwieldy that it is unsuitable for practical applications. We therefore have to develop a simpler method of estimating the dynamic characteristics of sensitive elements which, nevertheless, will take into account all the significant factors.

The above classification of forces and phenomena is also valid for the sensitive elements (membranes) of pressure transducers, and the same problems arise when an attempt is made to formulate a mathematical description of the process. If we neglect damping, the differential equation of motion of a circular membrane of constant thickness can be written in the following form in polar coordinates [57]

$$\Delta\Delta z + \frac{q}{D}\frac{\partial^2 z}{\partial \tau^2} = \frac{1}{D} P(r, \phi, \tau)$$

where z is the displacement from the position of equilibrium, $q = \rho h$ is the mass per unit surface area of the membrane, ρ is the density of the material, and h is the thickness of the membrane.

The cylindrical stiffness of the membrane is given by

$$D = Eh^3/[12(1 - \mu^2)] \tag{46}$$

In terms of the polar coordinates r and ϕ, the Laplace operator is

$$\Delta = \frac{\partial^2}{\partial r^2} + \frac{1}{r}\frac{\partial}{\partial r} + \frac{1}{r^2}\frac{\partial^2}{\partial \phi^2}$$

If we consider a circular membrane of radius r_0, which is clamped along the periphery, we have $r = r_0$

$$z = 0, \quad \partial z/\partial r = 0$$

Equation (45) acquires an additional term of the form $\pm N\Delta z/D$ when the clamped periphery of the membrane is uniformly stretched or compressed. The upper sign corresponds to compression and the lower to extension (N is the force per unit length of the membrane periphery).

Substituting $P = 0$ into the right-hand sides of (44) and (45), we obtain the equations for the free oscillations of the sensitive element, and the solutions of these equations can be used to obtain the natural (resonance) frequencies of the transducers.

The mathematical description of the oscillations of sensitive elements to which time-dependent forces are applied is thus seen to be relatively complicated. However, when we analyze the dynamic characteristics of velocity and pressure transducers, we do not require the entire range of information that is contained in the solutions of these differential equations. It is often enough to know the displacement at a particular point of the sensitive element, and only one value in the infinite discrete spectrum of resonance frequencies (the minimum natural frequency). In this sense, the sensitive elements of velocity and pressure transducers are similar to oscillatory systems with one degree of freedom.

13. OSCILLATIONS WITH A SYSTEM OF ONE DEGREE OF FREEDOM

The oscillations of a linear system with one degree of freedom are the simplest case of oscillatory motion. This motion is performed, for example, by a load of mass m suspended from a weightless spring of stiffness c. When a time-dependent force $P(\tau)$ is applied to the load, and there is also a resistive force that is proportional to the velocity of the load, the differential equation for the motion of the center of gravity of the load is

$$m\ddot{y} + k\dot{y} + cy = P(\tau) \tag{47}$$

or

$$\ddot{y} + 2h\dot{y} + \omega_r^2 y = P(\tau)/m \tag{47a}$$

where $2h = k/m$ and $\omega_r^2 = c/m$ is the square of the resonance angular frequency.

Free oscillations with damping. Let us begin by considering free oscillations, for which $P(\tau) = 0$. Substituting $y = u(\tau)\exp(-h\tau)$ in (47a), we obtain

$$\ddot{u} + (\omega_r^2 - h^2)u = 0$$

The solution of this equation is found to depend on the value of the difference $\omega_r^2 - h^2$. There are three possible cases.

Case 1. $\omega_r^2 - h^2 > 0$ (moderate damping). The solution is

$$u = A\cos\omega_1\tau + B\sin\omega_1\tau$$

$$y = (A\cos\omega_1\tau + B\sin\omega_1\tau)\exp(-h\tau)$$

or

$$y = A_1\sin(\omega_1\tau + \phi)\exp(-h\tau) \tag{48a}$$

where $\omega_1^2 = \omega_r^2 - h^2$ and A_1 and ϕ are the initial values of the amplitude and the phase.

It is clear from (48a) that the amplitude of the oscillations decreases exponentially with time. The oscillation period is given by

$$T = (2\pi/\omega_1) \simeq (2\pi/\omega_r)\left(1 + h^2/2\omega_r^2\right) = T_0\left(1 + h^2/2\omega_r^2\right)$$

where T_0 is the period of natural oscillations in the absence of damping. For example, when $h/\omega_r = 0.1$ we have $T = 1.005\,T_0$, i.e., the change in the period of natural oscillations is small when the damping is moderate.

Case 2. $\omega_r^2 - h^2 < 0$, $\omega_2^2 = h^2 - \omega_r^2$, i.e., this is the case of "strong" damping. The solution of the above equation is then expressed in terms of hyperbolic functions:

$$y = (A\cosh\omega_2\tau + B\sinh\omega_2\tau)\exp(-h\tau) \tag{48b}$$

The motion is now aperiodic and the deflection y decreases rapidly with time. The position of equilibrium can be approached both for $y > 0$ and $y < 0$, depending on ω_2.

Case 3. $\omega_r^2 - h^2 = 0$. This is the case of *critical damping*. The solution is

$$y = (A\tau + B)\exp(-h\tau) \tag{48c}$$

Forced oscillations with damping. We now assume that the external force is harmonic:

$$\frac{1}{m}P(\tau) = H\sin\omega\tau$$

The equation of motion that follows from (47a)

$$\ddot{y} + 2h\dot{y} + \omega_r^2 y = H\sin\omega\tau \tag{49}$$

The solution of this inhomogeneous equation can be sought in the form of the sum of the general solution y_1 of the homogeneous equation and a particular solution y_2 of the inhomogeneous equation. It was shown above [see (48)] that the form of y_1 depends on the difference $\omega_r^2 - h^2$ and contains the factor $\exp(-h\tau)$. Hence, $y_1 \neq 0$ only during the very short initial period of time following the application of the external force, i.e., the free oscillations are rapidly damped out and only the forced oscillations y_2 remain.

Let us now take this special solution to be

$$y_2 = A \sin(\omega\tau + \delta)$$

where A and δ are constants independent of time.

Substituting for y_2 in the initial equation (49), and collecting together the coefficients of $\sin \omega\tau$ and $\cos \omega\tau$, we obtain

$$-A\omega^2 \sin \delta - 2Ah\omega \sin \delta + A\omega_r^2 \cos \delta = H$$

$$-A\omega^2 \sin \delta + 2Ah\omega \cos \delta + A\omega_r^2 \sin \delta = 0$$

Simple algebra then eventually yields or

$$-2Ah\omega = H \sin \delta$$

$$A(\omega_r^2 - \omega^2) = H \cos \delta$$

$$\tan \delta = -2hH\omega/(\omega_r^2 - \omega^2)$$

$$A = H/\sqrt{(\omega_r^2 - \omega^2)^2 + 4h^2\omega^2}$$

The static displacement of the spring when the force P is applied to it is

$$y_{st} = P/c = mH/c = H/\omega_r^2$$

The dynamic increase in the amplitude of the oscillations is

$$\lambda = A/y_{st} = 1/\sqrt{(1-\bar{\omega}^2)^2 + \alpha^2 \bar{\omega}^{-2}} \qquad (50)$$

where $\bar{\omega} = \omega/\omega_r$ and $\alpha = 2h/\omega_r$.

The phase shift δ between the sinusoidal applied force and the displacement y is given by

$$\sin \delta = -\alpha \bar{\omega} \lambda \qquad (51)$$

Thus, the harmonic external force acting on the linear oscillatory system produces a harmonic steady-state response after a certain interval of time has elapsed, and the amplitude and phase of this response is given by (50) and (51). The free oscillations are rapidly damped out, and the values of λ and δ depend both on the reduced frequency $\bar{\omega}$ and on the damping factor α (Fig. 50).

For moderate damping ($\alpha < 1.4$), the function $\lambda(\bar{\omega})$ has a more or less well defined resonance maximum

$$\lambda_{max} = (1 - \alpha^2/4)^{-1/2}/\alpha$$

at $\bar{\omega} = (1 - \alpha^2/2)^{1/2}$.

For weak damping ($\alpha \ll 1$), the maximum resonance value of λ is $1/\alpha$, and this occurs at $\bar{\omega} \simeq 1$. The maximum value λ_{max} decreases with increasing α, and its position on the frequency scale shifts toward lower values.

In the case of critical damping ($\alpha = \sqrt{2} \simeq 1.4$), the maximum value of λ occurs at $\bar{\omega} = 0$, i.e., $\lambda(\bar{\omega}) \leqslant 1$ at all frequencies.

106 DYNAMIC CHARACTERISTICS OF TRANSDUCERS

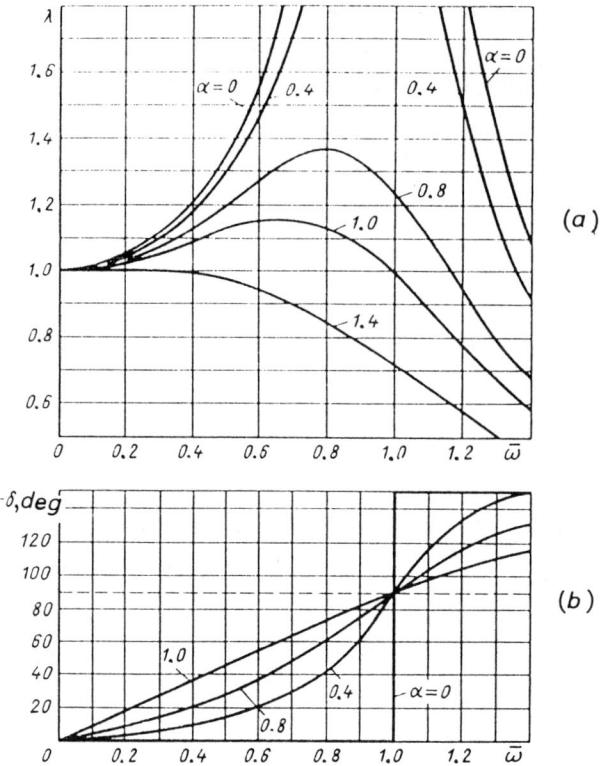

Fig.50. Amplitude–frequency (a) and phase–frequency (b) characteristics of a system at one degree of freedom

The amplitude–frequency characteristic of a linear oscillatory system with one degree of freedom is thus determined by a single parameter, namely, the damping coefficient α. The phase–frequency characteristic $\delta(\bar{\omega})$ is also completely determined by the value of α [see (51)]. When $\alpha = 0$, the phase shift δ changes discontinuously from zero to π at $\bar{\omega} = 1$. When $\alpha > 0$, the variation of δ with frequency is smoother. Whatever the value of α, we then have $\delta = \pi/2$ at $\bar{\omega} = 1$. Damping forces the oscillations to lag in phase behind the external force ($\delta < 0$).

Since the amplitude–phase and phase–frequency characteristics of a linear system with one degree of freedom are readily calculated, it is interesting to consider whether the sensitive elements of velocity and pressure transducers can be represented by some equivalent oscillatory systems with one degree of freedom. The determination of the dynamic characteristics can then be reduced to the determination of the values of ω_r and α. This approach is adopted below.

14. DYNAMIC CHARACTERISTICS OF VELOCITY TRANSDUCERS FOR TRANSPARENT LIQUIDS [19]

In Chapter 3, we examined optical-fiber velocity transducers for transparent fluids. A typical arrangement of this kind is shown in Fig. 51a (see also Fig. 27). The sensitive element of the transducer *1* is partially shielded, so that only the portion of length b is present in the flow. The deflection is recorded at $x = b$.

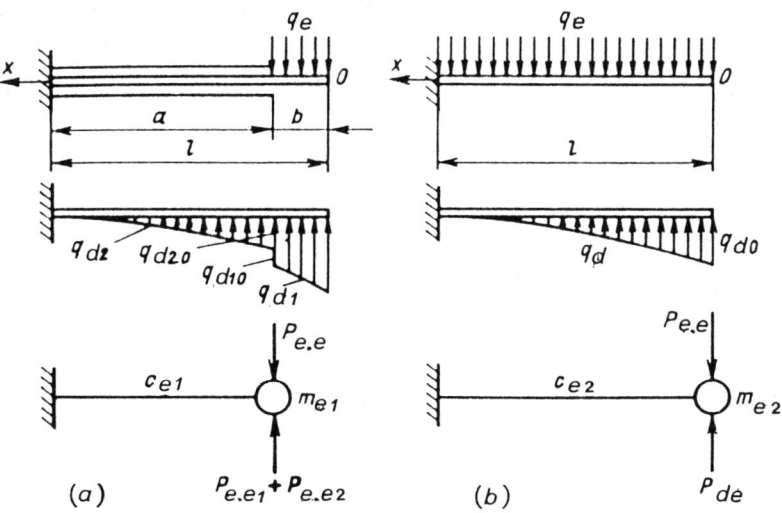

Fig. 51. Forces acting on sensitive elements in a flow

The other arrangement (velocity transducer 2) is shown in Fig. 51b (see Figs. 28 - 30). Here the elastic sensitive element is a "bare" cantilever segment of an optical fiber, and the deflection is recorded at $x = 0$.

It is thus clear that the velocity-measuring system records the oscillations of the sensitive element at a single point. It is therefore useful to replace the sensitive element with an equivalent system with one degree of freedom, and then use the simple theory given in the last section to calculate the dynamic characteristics of the velocity transducer. This approach is used in the theory of oscillations to determine the resonance frequencies of rods with localized masses [57]. It is convenient to assume that the rod is weightless, and to replace its distributed mass by some equivalent mass localized at a convenient point. The equivalent mass is chosen so that the resonance frequency of the weightless rod with the localized equivalent mass is the same as that of the original massive rod. It will be shown below that an analogous approach is possible in the case of damped forced oscillations of a sensitive element under distributed forces. The latter can be replaced by equivalent forces localized at the point of detection. These equivalent forces are determined from the condition that the recorded deflection is the same as for the actual external and damping forces.

It follows that, if we are to replace the sensitive element by an equivalent oscillatory system with one degree of freedom, we must determine the minimum resonance frequency of the sensitive element and the forces acting upon it.

Resonance frequency of sensitive element. If we neglect damping in (44) and put $P(\tau) = 0$, we obtain the following equation for the free oscillations of a weightless rod:

$$EI \frac{\partial^4 y}{\partial x^4} + \rho F \frac{\partial^2 y}{\partial \tau^2} = 0 \tag{52}$$

If we introduce the dimensionless coordinate $\xi = x/l$, and rearrange the equation we obtain

$$\frac{\partial^4 y}{\partial \xi^4} + \frac{pFl^4}{EI} \frac{\partial^2 y}{\partial \tau^2} = 0 \tag{52a}$$

The solution is [57]

$$y(\xi, \tau) = Y(\xi)\,(C_1 \cos \omega\tau + C_2 \sin \omega\tau)$$

Substituting this solution into the initial equation, we obtain an ordinary differential equation for $Y(\xi)$:

$$\frac{d^4 Y}{d\xi^4} - r^4\, Y(\xi) = 0 \tag{53}$$

where

$$r^4 = \rho F \omega^2 l^4 / EI \tag{54}$$

The solution of [53] is

$$Y(\xi) = A \cosh r\xi + B \sinh r\xi + C \cos r\xi + D \sin r\xi \tag{55}$$

and can be written in the form

$$Y(\xi) = A\,Y_1(\xi) + B\,Y_2(\xi) + C\,Y_3(\xi) + D\,Y_4(\xi) \tag{55a}$$

where

$$Y_1(\xi) = \frac{1}{2}\,(\cosh r\xi + \cos r\xi)$$

$$Y_2(\xi) = \frac{1}{2r}\,(\sinh r\xi + \sin r\xi)$$

$$Y_3(\xi) = \frac{1}{2r^2} (\cosh r\xi - \cos r\xi)$$

$$Y_4(\xi) = \frac{1}{2r^3} (\sinh r\xi - \sin r\xi) \tag{56}$$

The $Y_i(\xi)$ are usually referred to as the *Krylov functions*. The constants A, B, C, and D are determined from the boundary conditions. For a cantilever clamped at $\xi = 1$, the boundary conditions are

$$\text{at } \xi = 0 \quad Y'' = Y''' = 0$$

$$\text{at } \xi = 1 \quad Y = Y' = 0 \tag{57}$$

It is more convenient to take the solution in the form given by (55a). The initial conditions (57) then give C, D = 0, and

$$A Y_1(1) + B Y_2(1) = 0$$

$$A Y_1'(1) + B Y_2'(1) = 0$$

Since these two inhomogeneous linear equations with two unknowns A and B must be compatible, we have

$$Y_1(1) Y_2'(1) - Y_1'(1) Y_2(1) = 0$$

Substituting for Y_1 and Y_2, and rearranging, we obtain the transcendental characteristic equation, $\cosh r \cos r = 1$, which has an infinite number of solutions. The first two roots of this equation are $r_1 = 1.8751$ and $r_2 = 4.6941$. The only interesting root is r_1, which corresponds to the fundamental frequency of the cantilever (minimum natural frequency). According to (54), the angular frequency is given by

$$\omega_{r1} = (r_1^2/l^2)(EI/\rho F)^{1/2} \tag{58}$$

The angular frequency of the cantilever is

$$f_{r1} = \omega_{r1}/2\pi \tag{58a}$$

The last two expressions can be used to calculate the resonance frequencies of cantilever sensitive elements in velocity transducers for transparent media (Chapter 3).

Calculations based on (58) are found to be in good agreement with the measured resonance frequencies of sensitive elements of actual velocity transducers. Such measurements are performed by two methods that differ in the method employed to excite the oscillations of the sensitive element. The simplest method relies on the acoustic excitation of the oscillations by a suitable vibrator. By varying the frequency, and measuring the transducer signal on the oscillograph screen, it is possible to determine the resonance frequency from the position of the signal maximum. However, it is important to ensure that the increase in the signal is not due to a resonance peak on the amplitude–frequency characteristic of the vibrator itself. Figure 52 shows the signal amplitude as a function of frequency for the velocity transducer shown in Fig. 28. It is clear that the resonance frequency of the sensitive element is 665 Hz. However, this curve cannot be regarded as the amplitude–frequency characteristic of the velocity transducer because the amplitude of the exciting force applied by the acoustic vibrator depends on frequency. In particular, this force is always zero for $f = 0$.

The amplitude–frequency characteristic can also be obtained by exciting the oscillations in another way, in which the amplitude of the exciting force is independent of frequency. A thin metal layer is deposited on the sensitive element and on the transducer lightguides of Fig. 27a. One of the terminals of an acoustic oscillator is connected to the conducting coating of the sensitive element, and the other is grounded. A potential of about 1000 V relative to ground is applied to the lightguide coating. The amplitude of the transducer signal is then measured as a function of the

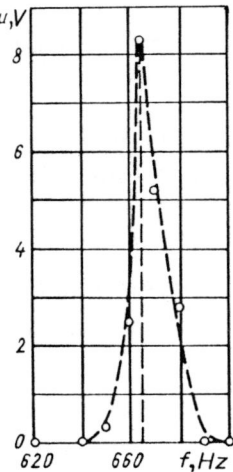

Fig.52. Resonance curve of the velocity transducer shown in Fig.28

frequency of the acoustic oscillator. The resulting curve (points *1* in Fig. 53) is the amplitude–frequency characteristic of the velocity transducer because the amplitude of the applied force is determined only by the potential applied to the lightguides and the sensitive element, and is independent of frequency. Curve *2* in Fig. 53 was calculated from (50) for $\alpha = 0.02$. It is clear that there is good agreement between experiment and calculations, indicating that the sensitive element can indeed be replaced by an equivalent oscillatory system with one degree of freedom. One of the parameters of the equivalent system with one degree of freedom, i.e., its resonance frequency, can be both calculated and measured.

Equivalent stiffness and mass. The resonance frequency of the sensitive element of the velocity transducers discussed above can be determined from (58) and (58a). The equivalent stiffness of the sensitive element is given by

$$c_{el} = 3EI/a^3 \tag{59}$$

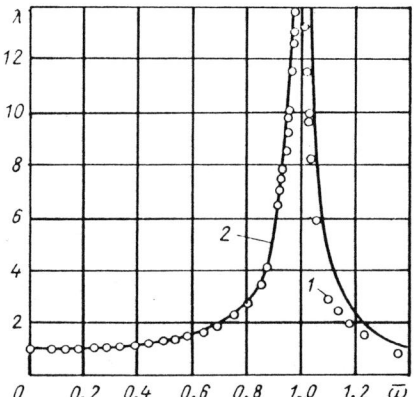

Fig.53. Amplitude–frequency characteristic of a velocity transducer

$$c_{e2} = 3EI/l^3$$

Hence, the equivalent masses are given by

$$m_{e1} = c_{e1}/\omega_{r1}^2 \simeq 0.243\, \rho F l^4/\alpha^3 \qquad (60)$$

$$m_{e2} = c_{e2}/\omega_{r1}^2 \simeq 0.243 \rho F l \qquad (61)$$

Equivalent forces and damping coefficients. To calculate the equivalent forces, we need the equation for the elastic line $y(x)$ of the sensitive element loaded as shown in Fig. 51.

For transducer 1, we have [41]

$$0 < \bar{x} < b: \quad y(\bar{x}) = (q_e l^4/24EI)\,[3 - 4\bar{a}^{-3} + \bar{a}^{-4} - 4(1-\bar{a}^{-3})\bar{x} + \bar{x}^{-4}]$$

$$\bar{b} < \bar{x} < 1: \quad y(\bar{x}) = (q_e l^4/24EI)\,[3 - 4\bar{a}^{-3} + \bar{a}^{-4} - 4(1-\bar{a}^{-3})\bar{x} + \bar{x}^{-4} - (\bar{x}-\bar{b})^4]$$

$$\bar{x} = x/l, \quad \bar{a} = a/l, \quad \bar{b} = b/l$$

Since the deflection is determined at the point $\bar{x} = \bar{b} = 1 - \bar{a}$, transducer *1* can be optimized using the requirement of maximum sensitivity to velocity. The point is that this sensitivity is zero for $\bar{a} = 0$ and $\bar{a} = 1$. Consequently, there is an optimum value of $\bar{a}$ for which the sensitivity is a maximum. At the point $\bar{x} = \bar{b} = 1 - \bar{a}$, the deflection is

$$y(\bar{b}) = (q_e l'/24EI)\,[3 - 4\bar{a}^{-3} + \bar{a}^{-4} - 4(1 - \bar{a}^{-3})(1 - \bar{a}) + (1 - \bar{a})^4]$$

Differentiating this equation with respect to $\bar{a}$, and equating the derivative to zero, we obtain the equation for the optimum value of $\bar{a}$. The solution of this gives $\bar{a} = 0.686$. For the sake of simplicity, we may take $\bar{a} = 0.7$ ($\bar{b} = 0.3$), since the maximum value $y_{max}(\bar{b})$ is not well-defined, i..e, a 10% change in $\bar{a}$ in the region of the maximum produces a 1% change in $y_{max}(\bar{b})$. The equations of the elastic line for $\bar{a} = 0.7$ are as follows:

$$0 < \bar{x} < 0.3: \quad y(\bar{x}) = (q_e l^4/24EI)\,(1.868 - 2.628\,\bar{x} + \bar{x}^{-4})$$

$$0.3 < x < 1: \quad y(\bar{x}) = (q_e l^4/24EI)\,(1.860 - 2.520\bar{x} - 0.540\bar{x}^{-2} + 1.200\bar{x}^{-3})$$

At $\bar{x} = 0.3$, the deflection is

$$y_{01} = 1.088\ q_e l^4/24EI \tag{62}$$

For transducer *2*, we have

$$y(\bar{x}) = q_e l^4\,(3 - 4\bar{x} + \bar{x}^{-4})/24EI$$

At $\bar{x} = 0$, the deflection is

$$y_{02} = q_e l^4/8EI \tag{63}$$

From now on, we shall assume that the external load q_e has a harmonic time component which forces the sensitive element into harmonic oscillations. The velocity $\dot{y}(x)$ is then a function of position along the sensitive element, and is proportional to $y(x)$.

For transducer *1*, we have

$$0 < \bar{x} < 0.3: \quad \dot{y}(\bar{x})/\dot{y}_{01} = 1.717 - 2.415\bar{x} + 0.919\bar{x}^{-4}$$

$$0.3 < \bar{x} < 1: \quad \dot{y}(\bar{x})/\dot{y}_{01} = 1.710 - 2.316\bar{x} - 0.496\bar{x}^{-2} + 1.103\bar{x}^{-3}$$

where $\dot{y}_{01}$ is the value of $\dot{y}$ at $\bar{x} = 0.3$.

For transducer *2*, we have

$$\dot{y}(\bar{x})/\dot{y}_{02} = 1 - \frac{4}{3}\bar{x} + \frac{1}{3}\bar{x}^{-4}$$

where $\dot{y}_{02}$ is the value of $\dot{y}$ at $\bar{x} = 0$.

In the linear approximation, the damping load is $q_d \simeq \dot{y}$, and its distribution along the length of the sensitive element is described by the following equations.

For transducer *1*,

$$0 < \bar{x} < 0.3: \quad q_{d1} = q_{d10}(1.717 - 2.415\,\bar{x} + 0.919\bar{x}^{-4}) \tag{64}$$

$$0.3 < \bar{x} < 1: \quad q_{d1} = q_{d20}(1.710 - 2.316\bar{x} - 0.496\bar{x}^{-2} + 1.103\bar{x}^{-3}) \tag{65}$$

and for transducer *2*,

$$q_d = q_{d0}\left(1 - \frac{4}{3}\bar{x} + \frac{1}{3}\bar{x}^{-4}\right) \tag{66}$$

where $q_{di\,0}$ is the value of q_{di} at the point $\bar{x} = 0.3$, and q_{d0} is the value of q_d at the point $\bar{x} = 0$.

The physical significance of the damping load q_{d2} is clear: it is due to viscous friction in the liquid filling the shields surrounding the element. The quantities q_{d1} and q_d require some further explanation.

Assuming that the velocity fluctuations are small in comparison with the average velocity $(w'/<w> \ll 1)$, we obtain the following expression in the case of a stationary sensitive element:

$$q'_e = w' \cdot \left.\frac{\partial q_e}{\partial w}\right|_{w = <w>}$$

If the sensitive element moves with velocity $\dot{y}$, then

$$q'_e = (w' - \dot{y}) \left.\frac{\partial q_e}{\partial w}\right|_{w = <w>}$$

Consequently, the part of the external load given by

$$\dot{y} \left.\frac{\partial q_e}{\partial w}\right|_{w = <w>}$$

which we denote by q_d, damps the oscillations of the sensitive element.

The equivalent damping force can now be determined by calculating the deflection of the sensitive element under the damping load on transducers *1* and *2* at the points $\bar{x} = 0.3$ and $\bar{x} = 0$, respectively. The first step is to find the equation of the elastic line by solving the differential equation

$$M_d(\bar{x})/EI = y''_d(\bar{x})$$

where $M_d(\bar{x})$ is the bending moment due to the damping load. Hence, for transducer *1* and $0.3 < \bar{x} < 1$,

$$M_{d1}(\bar{x}) = l^2 \int_0^{0.3} q_{d1}(\bar{\xi})\,(\bar{x} - \bar{\xi})\,d\bar{\xi}$$

$$M_{d2}(\bar{x}) = l^2 \int_{0.3}^{\bar{x}} q_{d1}(\bar{\xi})\,(\bar{x} - \bar{\xi})\,d\bar{\xi}$$

$$y'_{d1}(x) = l^3 \left[\int \frac{M_{d1}(\bar{x})}{EI}\,d\bar{x} + C_1 \right]$$

$$y_{d1}(x) = l^4 \int \left[\int \frac{M_{d1}(\bar{x})}{EI}\,d\bar{x} + C_1 \right] d\bar{x} + C_2$$

where $\bar{\xi}$ is the integration variable.

The quantities $y'_{d2}(\bar{x})$ and $y_{d2}(\bar{x})$ are found in a similar way. The integration constants are obtained from boundary conditions. For $\bar{x} = 1$:

$$y'_{d1} = y'_{d2} = y_{d1} = y_{d2} = 0$$

so that the deflection at $\bar{x} = 0.3$ is given by

$$y_{d10} = y_{d1}(0.3) = 0.063\, q_{d10}\, l^4 / EI \tag{67}$$

$$y_{d20} = y_{d2}(0.3) = 0.018\, q_{d20}\, l^4 / EI \tag{68}$$

The equivalent localized forces $P_{d.ei}$ are applied at $\bar{x} = 0.3$ and can be calculated from the condition that the deflections due $P_{d.ei}$ and q_{di} are equal:

$$P_{d.e1}\,(0.7l)^3 / 3EI = y_{d10} = 0.063\, q_{d10}\, l^4 / EI$$

$$P_{d.e1} = 0.547\, q_{d10}\, l = k_1\, \dot{y}_0 \tag{69}$$

$$P_{d.e2}\,(0.7l)^3 / 3EI = y_{d20} = 0.018\, q_{d20}\, l^4 / EI$$

$$P_{d.e2} = 0.161 q_{d20} \, l = k_2 \, \dot{y}_0 \tag{70}$$

For transducer *2*,

$$M_d(\bar{x}) = l^2 \int_0^{\bar{x}} q_d(\bar{\xi}) (\bar{x} - \bar{\xi}) \, d\bar{\xi}$$

The equivalent damping force is

$$P_{d.e} = 0.246 q_{d0} \, l = k \, \dot{y}_0 \tag{71}$$

The equivalent external force $P_{e.e}$ is found from the condition that the deflections due to q_e and $P_{e.e}$ are equal. Using (62) and (63), we find that for transducer *1*

$$P_{e.e1} (0.7l)^3 / 3El = 1.088 q'_e l^4 / 24 EI$$

$$P_{e.e1} = 0.396 q'_e \, l = 0.396 lw' \left. \frac{\partial q_e}{\partial w} \right|_{w = <w>} \tag{72}$$

whereas for transducer *2*

$$P_{e.e2} \, l^3 / 3 EI = q'_e l^4 / 8 EI$$

$$P_{e.e2} = 0.375 \, q'_e \, l = 0.375 lw' \left. \frac{\partial q_e}{\partial w} \right|_{w = <w>} \tag{73}$$

The equations of a variant of the equivalent system with one degree of freedom for velocity transducers *1* and *2* are:

$$m_{e1} \, \ddot{y}_0 + (k_1 + k_2) \, \dot{y}_0 + c_{e1} \, y_0 = P_{e.e1}$$

$$m_{e2}\ddot{y}_0 + k\,\dot{y}_0 + c_{e2}\,y_0 = P_{e.e2} \tag{74}$$

If we suppose that $\dot{y}_0 \ll <w>$, we find from (15) that the damping loads are

$$q_{d10} = q_{d0} = 0.5\rho_l d_{se}\,\dot{v}_0\,\partial(w^2 C_d)/\partial w \tag{75}$$

$$q_{d20} = 0.5\rho_l d_{se}\,C_d\,\dot{y}_0^2 \simeq 0.5\rho_l \nu_l A\,\dot{y}_0 \tag{76}$$

In the present case, the drag coefficient of the sensitive element oscillating in the liquid filling the shield can be represented by the approximate linear equation

$$C_d = A/\mathrm{Re}_d = A\nu_l/(d_{se}\,\dot{y}_0) \tag{77}$$

where A is a constant.

Estimates show that, in this case, Re_d is of the order of 10^{-2}–10^{-4} and the contribution of q_{d2} to the total damping does not exceed 10%. It follows that the use of this approximation does not introduce an appreciable error into the damping coefficient

$$\alpha_i = k_i/m_{ei}\,\omega_r \tag{78}$$

The value of ω_r is determined by the stiffness of the sensitive element, which, in turn, depends on the maximum velocity of the liquid that is measured by the transducer. Assuming that the maximum deflection of the sensitive element is equal to its diameter, and using (62) and (63), we obtain the following expressions for transducers 1 and 2, respectively:

$$d_{se} = 1.088\,\frac{0.5\,\rho_l d_{se}\,\omega_{max}^2\,C_{d\,max}\,l^4}{24 EI}$$

$$d_{se} = \frac{0.5\, \rho_l\, d_{se}\, \omega_{max}^2\, C_{d\,max}\, l^4}{8EI}$$

where $C_{d\,max}$ is the value of C_d at $w = w_{max}$.

These expressions can be used to determine the values of l, and then ω_r can be found from (58). Using (60), (61), (69) and (71), (75), (76) and rearranging, we obtain the following expressions for transducers *1* and *2*, respectively:

$$\alpha_1 = 0.82 \left(\rho_l / C_{d\,max}\, \rho_{se}\right)^{1/2} \left[\frac{1}{w_{max}} \frac{\partial(w^2 C_d)}{\partial w} + 0.29 \frac{A}{\mathrm{Re}_{max}} \right] \qquad (79)$$

$$\alpha_2 = 0.65 \left(\rho_l / C_{d\,max}\, \rho_{se}\right)^{1/2} \frac{1}{w_{max}} \frac{\partial(w^2 C_d)}{\partial w} \qquad (80)$$

The damping coefficient is thus seen to depend on the velocity of the liquid, the density ratio of the liquid and the material of the sensitive element, the upper limit for the measured velocity w_{max}, and the drag coefficient at this velocity, $C_{d\,max}$.

The values of C_d and $C_{d\,max}$ can be determined as functions of w and w_{max}, and of the properties of the liquid (ρ_l and ν_l), using (11), (13), and (14).

Figures 54a and b show the value of α calculated from (79) and (80) for air and water. It was assumed in these calculations that $\rho_{se} = 2.6 \times 10^3$ kg/m^3 (glass) and $A = 3.3$. For this value of A equation (77) approximates (11) to within $\pm 40\%$ for $\mathrm{Re}_d = 10^{-4} - 10^{-2}$, which introduces an error of only a few percent in α. It is clear from the figure that the damping coefficients are small even for the minimum values of the diameter of the sensitive element. For air, $\alpha = 0.01 - 0.1$ which means that simple electronics can be used to correct the amplitude–frequency characteristic. When the "corrector" is designed, we can put $\alpha = 0$ since the value of λ in this range is not very dependent on α. In particular, when $\overline{\omega} = 0.5$, we have $\lambda_{\alpha=0} \simeq \lambda_{\alpha=0.1}$ $\simeq 1.33$. For water, the values of α approach the critical value of 1.4 which often means that the amplitude frequency characteristic need not in fact be corrected.

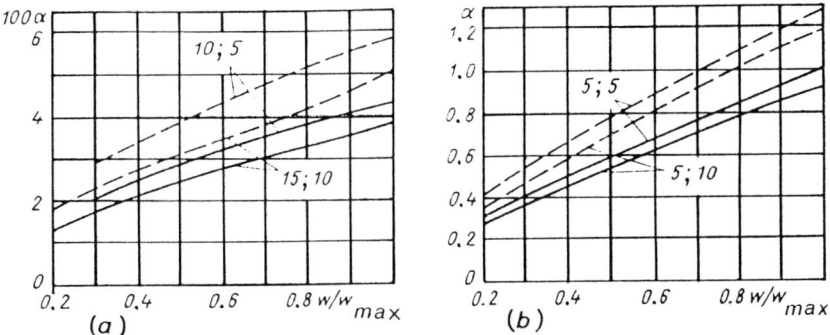

Fig.54 Damping coefficient of velocity transducers in air (a) and water (b): dashed curves— transducer No.1, solid curves — transducer No.2; numbers shown against curves: first — w_{max}, m/s, second — d_{se}, μm

Measurements of α in air are in good agreement with calculations. An example is shown in Fig. 53.

The resonance frequency of the sensitive element is an important dynamic parameter of velocity transducers. The resonance frequency f_r of the above transducer designs is shown in Fig. 55 as a function of the maximum velocity w_{max}. For air, f_r can reach 10 kHz. The resonance frequencies measured in water can be higher by an order of magnitude. As already noted, the dynamic characteristic of the optical–fiber velocity transducer in effect adjusts itself to the conditions of measurement. The higher the maximum measured velocity the stiffer the sensitive element, and the higher the resonance frequency. This means that the shift of the energy-bearing frequencies toward higher values in the velocity fluctuation spectrum is compensated by the increase in the natural frequency of the sensitive element.

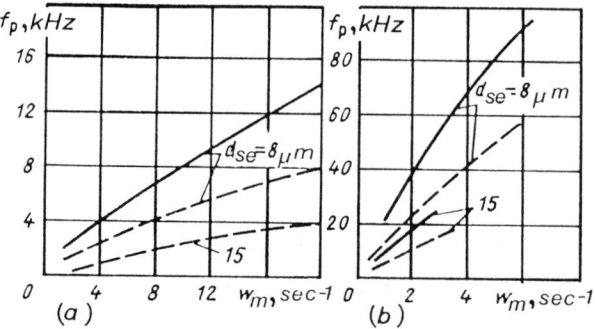

Fig. 55 Resonance frequencies of sensitive elements in air (a) and in water (b): dashed curves — transducer No. 1, solid curves — transducer No. 2

15. DYNAMIC CHARACTERISTICS OF VELOCITY TRANSDUCERS FOR OPAQUE LIQUIDS

The dynamic characteristics of a velocity transducer for opaque liquids are conveniently analyzed in terms of the equivalent oscillatory system with one degree of freedom by determining its resonance frequency and damping coefficient.

Resonance frequency. The velocity transducer for an opaque liquid is a complex oscillatory system. Its dynamic characteristics are determined by two oscillatory components, namely, the sensitive element and the pointer. The latter is fused into the end of the elastic sensitive element and can have somewhat lower resonance frequency than indicated by (58) in which it is assumed that one end of the cantilever is rigidly clamped. Experience gained by using velocity transducers in liquid metals suggests that the elastic clamping of one of the ends of the pointer has practically no effect on its resonance frequency, provided the pointer does not touch the inner wall of the sensitive element and is firmly fused to its end. Measurements of the resonance frequency of the pointer are therefore necessary not so much as a way of verifying the validity of calculations based on (58) as for verifying the fabrication quality of the velocity transducer.

On the other hand, the pointer constitutes a load for the sensitive element and must therefore affect the natural frequency of the latter. This can be estimated analytically by considering a cylindrical model of the sensitive element (Fig. 38a). It is clear that the calculated values of ω_r obtained for this model constitute the lower limit for this quantity. This estimate is of practical value in view of the difficulties in measuring the resonance frequency of sensitive elements of velocity transducers for opaque media (see below).

We must now consider the equation of motion of a hollow cylindrical cantilever, taking into account the forces and moments introduced by the pointer. We shall replace the pointer by a force Q applied to the end of the sensitive element. We shall suppose that the entire mass m_p of the pointer is localized at its center of gravity, which lies at a distance $L/2$ from the free end of the sensitive element. Taking the origin of coordinates $(x = 0)$ at the point at which the cantilever is clamped, we obtain the following equations for the force and its moment:

$$m_p \frac{\partial^2 y_{cm}}{\partial \tau^2} = - Q(l, \tau)$$

$$m_p \rho_i^2 \left.\frac{\partial^2 \theta(x, \tau)}{\partial \tau^2}\right|_{x=l} = \frac{Q(l, \tau) L}{2} - EI \left.\frac{\partial^2 y(x, \tau)}{\partial x^2}\right|_{x=l}$$

where $y_{cm}(\tau) = y(l, \tau) - \frac{1}{2} L\, \theta(l, \tau)$ is the displacement of the center of gravity of the pointer, $\rho_i = L/\sqrt{12}$ is the radius of inertia of the pointer about the axis passing through its center of gravity at right angles to the plane of the oscillations of the system, $\theta(l, \tau) = y'(x, \tau)|_{x=l}$ is the angle of rotation of the cross section of the cantilever at the free end of the sensitive element, E is Young's modulus, and I is the second moment of area of the cross section of the sensitive element.

Eliminating y_{cm} and $\theta(l, \tau)$ from these equations we obtain

$$EI \left.\frac{\partial^2 y(x, \tau)}{\partial x^3}\right|_{x=l} - m_p \left.\frac{\partial^2 y(x, \tau)}{\partial \tau^2}\right|_{x=l} + \frac{1}{2} m_p L \left.\frac{\partial^3 y(x, \tau)}{\partial \tau^2 \partial x}\right|_{x=l} = 0 \quad (81)$$

$$\frac{1}{2} EIL \left.\frac{\partial^3 y(x, \tau)}{\partial x^3}\right|_{x=l} - EI \left.\frac{\partial^2 y(x, \tau)}{\partial x^2}\right|_{x=l} - m_p \rho_i^2 \left.\frac{\partial^3 y(x, \tau)}{\partial \tau^2 \partial x}\right|_{x=l} = 0 \quad (82)$$

The equation of free oscillations given by (52a) must be solved subject to these boundary conditions. The solution is conveniently written in the form given by (55a). Using the boundary conditions $Y = Y' = 0$ at $\xi = 0$, we obtain $A = B = 0$. Therefore

$$y(\xi, \tau) = (C Y_3 + D Y_4)(c_1 \cos \omega \tau + c_2 \sin \omega \tau) = Y(\xi)\, \Phi(\tau)$$

Substituting for $y(\xi, \tau)$ in the boundary conditions given by (81) and (82), and bearing in mind the fact that $\ddot{\Phi}(\tau) = -\omega^2 \Phi(\tau)$, we obtain

$$Y'''(1) + \alpha r^4 Y(1) - \alpha \epsilon r^4 Y'(1) = 0$$

$$-\epsilon Y'''(1) + Y''(1) - \alpha \delta r^4 Y'(1) = 0$$

where $\alpha = m_p/m_{se}$, $\delta = \rho_i^2/l^2$, and $\epsilon = L/2l$.

Substituting $Y(1) = CY_3(1) + DY_4(1)$, we obtain the following two equations for the constants C and D:

$$[Y'''_3(1) + \alpha r^4 Y_3(1) + \alpha \epsilon r^4 Y'_3(1)] C + [\alpha r^4 Y_4(1) - \alpha \epsilon r^4 Y'_4(1)] D = 0$$

$$[-\epsilon Y'''_3(1) + Y''_3(1) - \alpha \delta r^4 Y'_3(1)] C +$$
$$+ [-\epsilon Y'''_4(1) + Y''_4(1) - \alpha \delta r^4 Y'_4(1)] D = 0$$

These equations are consistent if the determinant of the coefficients of the unknowns C and D is zero. After some laborious algebra, we finally find that

$$\frac{1}{\alpha}(1 + \cosh r \cos r) - r(\sin r \cosh r - \cos r \sinh r) + 2\epsilon r^2 \sin r \sinh r -$$

$$- (\delta + \epsilon^2)(\sin r \cosh r + \cos r \sinh r)r^3 + \alpha \delta r^4 (1 - \cos r \cosh r) = 0 \quad (83)$$

The roots of this equation, r_i, appear in the expression for the resonance frequencies, given by (58). The values of r_1 and $f_{r,se}$ for the above model, calculated from (83) and (58), are plotted in Fig. 56 for different values of the diameter and length of the sensitive element. It is clear from the Figure that the effect of the pointer is to reduce the resonance frequency of the cylindrical sensitive element. Thus, in the absence of the pointer, $r_1 = 1.875$, whereas the values obtained from (83) lie in the range $r_1 = 0.27 - 0.68$, depending on the dimensions of the sensitive element. Despite this, the values $f_{r,se} = 1.3 - 6.5$ kHz are relatively high and significantly exceed (by a factor of at least two) the values of $f_{r,p}$ for the pointer (670 Hz). Hence it follows that the dynamic characteristics of the velocity transducer are determined in practice by the amplitude–frequency characteristics of the pointer if the damping coefficient of the sensitive element is not very large.

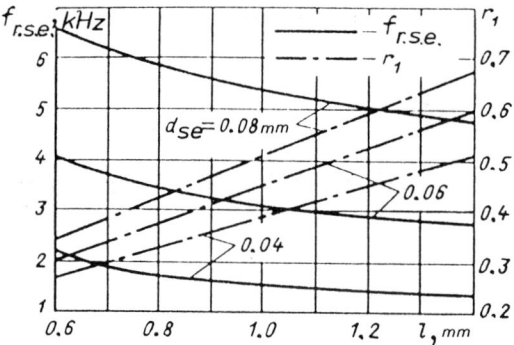

Fig.56 Calculated r_1 and $f_{r.se}$ ($L = 5$ mm, $d_y = 0.025$ mm, $\beta = 0.9$)

We note that we assumed that the pointer was perfectly rigid, but this is not, in fact, the case. Actually, (81) and (82) were obtained under the assumption that the displacement of the center of gravity of the pointer was directly related to the displacement of the tip of the sensitive element, whereas it follows from (50) that, when $\omega_{r.se} \gg \omega_{r.p}$ and $\omega = \omega_{r.se}$, it turns out that $\overline{\omega}_p \gg 1$ and, correspondingly, $\lambda_p \ll 1$. In other words, when $\omega_{r.se}/\omega_{r.p} \gg 1$, the center of gravity of the pointer remains at rest even when the amplitude of the resonance oscillations of the sensitive element is large. Accordingly, the effect of the pointer on $\omega_{r.se}$ will be much smaller than indicated by the above calculations.

It is clear from the above discussion that $f_{r.se}$ is quite difficult to measure. In particular, it cannot be measured by using an optical–fiber displacement transducer to determine the deflection of the pointer. Only one resonance peak is observed under acoustic excitation in all cases, and this occurs at a frequency close to the calculated value of f_r calculated for the pointer. This is so because $\lambda_p \ll 1$ for $f_{r.se}$, and the pointer does not react to the resonance oscillations of the sensitive element. These oscillations were therefore recorded with an additional displacement transducer. The tip of the sensitive element with the fused pointer was blackened and placed in the gap between the ends of illuminating and receiving lightguides. Several resonance peaks were observed as the frequency of the vibrator was varied. They corresponded to the resonances of the sensitive element, the system of measuring lightguides, the supporting structure, and so on. The required resonance was identified by depositing a drop of rapidly evaporating liquid (ethyl alcohol) on the sensitive element. If the particular resonance disappeared as a result of the deposition of the drop, and then reappeared as soon as the drop evaporated, this was an unambiguous indication that the particular peak represented a resonance of the sensitive element.

As expected, the resonance frequencies of sensitive elements measured in this way were found to be much higher than those calculated for the cylindrical model. For example, for a velocity transducer working in mercury and incorporating a sensitive element with tip diameter $d_{se} = 0.03$ mm and diameter–doubling length $l_2 \simeq 1.3$ mm (see Section 10), it was found that $f_{r.se} \simeq 8$ kHz. The pointer had a diameter of 0.025 mm and a length of 4.4 mm. Its resonance frequency was found to be 1 kHz which is in good agreement with (58a).

Equivalent forces and damping coefficients. The results obtained for the velocity transducer for transparent liquids (No.2) can be used to determine the equivalent damped and forced oscillations of sensitive elements. Thus, the values of $R_{d.e}$ and $R_{e.e}$ can be calculated from (71) and (73), and the damping coefficient of the sensitive element can be found from

$$\alpha_{se} = k\omega_r/c_e \qquad (84)$$

where c_e is determined from (59). The factor k is calculated from (71) and (75), and l is found from the condition that the displacement of the pointer $\overline{\Delta r_{max}} = \Delta r_{max}/r_r \sim 0.5 \sim d_p/(4r_r)$ at maximum velocity w_{max} lies on the linear segment of the characteristic of the displacement transducer (see Fig. 11).

If the transducer is to be used for measurements in a turbulent flow of a liquid metal, the value of w_{max} must be found from the condition

$$\mathrm{Re}_d = w_{max}\, d_{se}/\nu_l = 50$$

since the formation of the vortex sheet/stream behind the sensitive element will limit the value of w_{max} in this case (see Section 8).

For the chosen values of d_p, d_{se}, L and w_{max}, the length l of the sensitive element in (58) and (59) can be found from the relation

$$\Delta r_{max} = d_p/4 = q_{max}\, l^4 (4\overline{L} - 3)/24 EI \tag{85}$$

where $\overline{L} = L/l$ and

$$q_{max} = 0.5\, \rho_l\, C_{d\,max}\, w_{max}^2\, d_{se} \tag{86}$$

Substituting for k, ω_r, and c_e from (59), (71), and (75) in (84) we obtain

$$\alpha_{se} = 18.5\, f_{r.se}\, w\, d_p\, C_d\, (1 - 0.09\, C_d^{0.74})/[w_{max}^2\, C_{d\,max}\, (4\overline{L} - 3)] \tag{87}$$

It is clear that $\alpha \to \infty$ as $\overline{L} \to 3/4$, and this has to be explained. The point is that, according, to (85), the displacement of the pointer consists of two components with opposite signs. When $\overline{L} \to 3/4$, the tip of the pointer remains at rest, whatever the displacement of the sensitive element, and this is equivalent to an infinite damping coefficient.

Example. Determine the dynamic error of a velocity transducer when a measurement is made of the root–mean–quare velocity fluctuation in a turbulent flow of mercury ($\nu_l \simeq 10^{-7}$ m^2/s) when its mean velocity is $<w> = 10$ cm/s and the upper frequency limit of the energy spectrum of the velocity fluctuations is $f_{lim} = 200$ Hz.

Table 2

f, Hz	0	40	80	120	160	200
$\bar{\omega}_p$	0	0.04	0.08	0.12	0.16	0.20
λ_p	1.000	1.002	1.006	1.015	1.026	1.042
$\bar{\omega}_{se}$	0	0.005	0.010	0.015	0.020	0.025
λ_{se}	1.000	1.000	1.000	1.000	1.000	1.000
λ_p	1.000	1.002	1.006	1.015	1.026	1.042

We shall now estimate the error for the velocity transducer for which $f_{r.se}$ and $f_{r.p}$ were measured. In this case, $d_{se} = 0.03$ mm, $d_p = 0.025$ mm, $f_{r.se} = 8000$ Hz, $f_{r.p} = 10^3$ Hz, $w_{max} = 0.17$ m/s, and $\bar{L} = 4.4$. We determine the damping coefficient from (87), using (14) to calculate C_d and $C_{d\,max}$. The result is $\alpha_{se} \sim 0.87$. The damping coefficient of the pointer in air can be set equal to zero (see Section 14). The dynamic amplitude amplification coefficient of the velocity transducer is given by

$$\lambda_{tr} = \lambda_p \lambda_{se}$$

According to (50), when $\alpha = 0$ we have

$$\lambda_p = 1/|1 - \bar{\omega}_p^2|$$

$$\lambda_{se} = 1/\left[(1 - \bar{\omega}_{se}^2)^2 + \alpha^2 \bar{\omega}_{se}^2\right]^{1/2}$$

where $\bar{\omega}_p = f/f_{r.p}$ and $\bar{\omega}_{s.e} = f/f_{r.se}$. The values of λ_p, λ_{se}, and λ_{tr} are given in Table 2 for different values of frequency.

It is clear from the figure that the dynamic error associated with the sensitive element can be neglected and we can put $\lambda_{tr} \simeq \lambda_p$. Assuming that the spectral intensity of velocity fluctuations remains constant within the given frequency range, we obtain the following estimate for the maximum dynamic error in the measured root-mean-square velocity pulsations:

$$\delta_{max} = (1/\overline{\omega}_{lim}) \left[\int (1 - \overline{\omega}_p^2)^{-1} d\overline{\omega}_p \right] - 1 = \left(\ln \frac{1 + \overline{\omega}_{lim}}{1 - \overline{\omega}_{lim}} \right) / 2\overline{\omega}_{lim} - 1$$

where the integral is evaluated between 0 and $\overline{\omega}_{lim}$. Since $\overline{\omega}_{lim} = 0.2$, we obtain $\delta_{max} = 0.014$.

The dynamic error is small and there is little point in using a corrector for the amplitude-frequency characteristic.

16. DYNAMIC CHARACTERISTICS OF OPTICAL–FIBER PRESSURE TRANSDUCERS

Since pressure measurement involves the determination of the deflection of the membrane at a single point (near the center), the pressure transducer can also be looked upon as an oscillatory system with one degree of freedom. The dynamic characteristics are then completely determined by the resonance frequency ω_r of the membrane and the damping coefficient α.

The resonance frequency of a membrane can be determined either experimentally or by calculation. The damping coefficient can also be measured directly if the sensitive element of the pressure transducer is in direct contact with the medium in which the pressure is measured. The use of long pulse transmission lines between the point of measurement and the transducer may introduce strong damping and make the measurement of time-dependent pressure impossible.

Resonance frequency of a membrane and the damping coefficient. If we put $P(r, \phi, \tau) = 0$ in (45), we obtain the equation for the free oscillations of a circular membrane. The solution of this equation may be sought in the form

$$z(r, \phi, \tau) = Z(r, \phi)\,(A \cos \omega_r \tau + B \sin \omega_r \tau)$$

Substituting this solution into the original differential equation, we obtain

$$\Delta\Delta Z - \sigma^4 Z = 0$$

where

$$\sigma^4 = \rho h\, \omega_r^2 / D \tag{88}$$

This equation has a solution only for certain specific values of the dimensionless parameter $\sigma_{ns} r_0$, which form an infinite discrete spectrum [57]. The numbers n and s are respectively equal to the number of nodal diameters and nodal circles. The smallest value $\sigma_{00} r_0 = 3.19$ corresponds to the lowest mode of the membrane for which the oscillations are axially symmetric and for which there are no nodal circles. According to (88), the first resonance of the membrane occurs when the angular frequency is given by

$$\omega_{00r} = \sigma_{00}^2\,(D/\rho h)^{1/2} = (D/\rho h)^{1/2}/r_0^2$$

where D is given by (46).

The corresponding cyclic frequency is

$$f_{00r} = 10.21\,(D/\rho h)^{1/2}/2\pi r_0^2 \tag{89a}$$

The angular frequency ω_{00} is higher if the membrane is stretched when the pressure transducer is made. The effect of stretching on ω_{00r} can be judged from Fig. 57 which shows $\sigma_{00} r_0$ as a function of the stretching load N per unit perimeter length. It is clear that a similar effect on the resonance frequency may be produced by applying a pressure to the membrane. The influence of static pressure is taken into account in terms of the deflection of

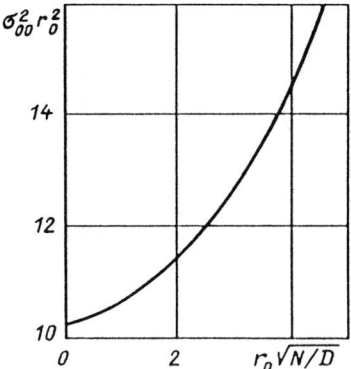

Fig.57 Resonance frequency as a function of the tensile load acting on a membrane

the membrane. The resonance formula can be calculated from the formula recommended in [47]:

$$f_{00r}^p = f_{00r}\left[1 + 1.464\, z_{01}^2/h^2\right]^{1/2}$$

where f_{00r} is given by (89a) and

$$z_{01} = pr_0^4/64D\left[1 + 0.0448\,(z_{01}/h)^2\right]$$

These equations are valid when $z_{01}/h < 1$.

When the pressure transducer is placed in a liquid of density ρ_l, some of the liquid oscillates together with the membrane, forming the so-called associated mass. We then have

$$f_{00r}^{\rho_l} = f_{00r}/(1+\beta)^{1/2}$$

$$\beta = 0.669\,(\rho_l/\rho)\,(r_0/h)$$

The simplest method of measuring the resonance frequency of a membrane is to excite it by acoustic radiation of known frequency, produced by a vibrator controlled by an acoustic oscillator (see Section 14).

Figure 58a shows a typical resonance curve obtained in this way. The resonance peak recorded for the transducer output in air is quite sharp so that the resonance frequency f_r can be measured to better than 1%. However, this method cannot be used to determine the amplitude-frequency characteristics of the transducer (see Section 14) because the amplitude of the harmonic external force is a function of frequency.

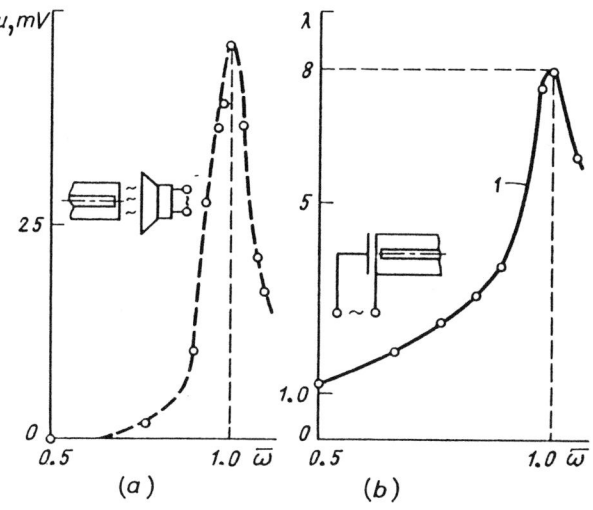

Fig.58 Resonance curve (a) and amplitude–frequency characteristic (b) of an optical–fiber pressure transducer

The amplitude-frequency characteristic is usually obtained by another common but more complicated method. The transducer membrane is covered by a thin layer of metal, evaporated on to it in a vacuum. A flat metal electrode is then placed parallel to the metallized membrane, and the alternating output of an acoustic oscillator is applied to the resulting plane parallel capacitor. The amplitude of the applied force is then determined only by the electrical potentials of the electrode and membrane, and is independent of frequency. The function $\lambda(\overline{\omega})$ obtained in this way (points in Fig. 58b) can be assumed to be the amplitude-frequency characteristic of the pressure transducer. The position of the peak can be used to determine the damping coefficient:

$$\alpha = 1/\lambda_{max} = 0.125$$

Curve *1* in Fig. 58b shows the calculated $\lambda(\alpha, \bar{\omega})$ obtained from (50) for the above value of α. It is in good agreement with the experimental points, showing that the pressure transducer can be modeled by an oscillatory system with one degree of freedom.

Once the amplitude–frequency characteristic and the value of α have been determined, the metal film can be removed, and f_r can be found by the acoustic excitation method described above. The amplitude–frequency characteristic can then be calculated for the new value of f_r, and the value of α for the metallized membrane can be found.

Measurements of the resonance frequencies have confirmed the theoretically-predicted excellent dynamic properties of thin glass membranes. For example, the resonance frequency of the membrane of the transducer whose calibration curve is shown in Fig. 22c was found to be 30 kHz, and the highly sensitive transducer used to measure velocity in a turbulent jet (see Section 7) was found to have $f_r \simeq 4$ kHz.

In contrast to velocity transducers whose damping coefficients depend on the flow rate, the dynamic characteristics of a pressure transducer can be obtained purely experimentally by the method described above. However, when relative long pulse transmission lines (capillaries) are present between the transducer and the point at which the pressure is measured, the additional damping due to these lines must be taken into account.

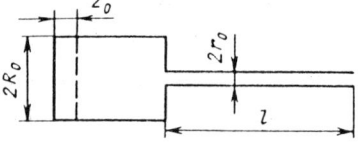

Fig.59 Model of a pressure transducer with a pulse line, used in the computation

Effect of pulse transmission lines. Consider the model of a pressure transducer illustrated in Fig. 59. The mass of the membrane can be neglected and we may assume that the equivalent mass is equal to the associated mass of the liquid that oscillates together with the membrane. The latter can be replaced by a weightless piston, held in position by an elastic force. The equivalent stiffness is given by

$$c_e = P\pi R_0^2/z_0 = P_{max}\pi R_0^2/z_{0\ max} \qquad (90)$$

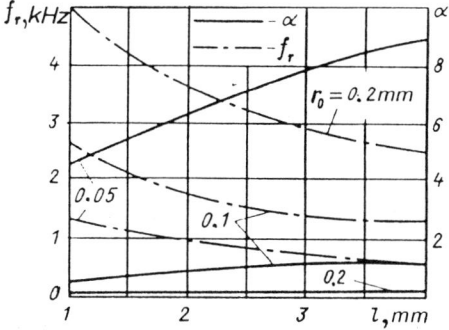

Fig.60 Resonance frequency and damping coefficient of a pressure transducer with a pulse line

where p is the pressure in the transducer cavity and z_0 is the displacement equal to the displacement of the membrane at its center.

In this approach, the resonance frequency

$$f_r = (c_e/m_e)^{1/2}/2\pi \qquad (91)$$

is independent of the mass of the membrane.

The kinetic energy of an elementary ring volume of the liquid in the capillary is

$$dT = \tfrac{1}{2}(2\pi r l)\, \rho_l w^2(r)\, dr \qquad (92)$$

Assuming that the radial velocity distribution is

$$w(r) = 2\overline{w}\,(1 - r^2/r_0^2) \tag{93}$$

and integrating (92) between zero and r_0, we obtain

$$T = \tfrac{2}{3}\pi l r_0^2 \rho_l \,\overline{w}^2 = \tfrac{1}{2} m_e \dot{z}_0^2$$

where

$$\overline{w} = (\pi R_0^2/\pi r_0^2)\,dz_0/d\tau = R_0^2 \dot{z}_0/r_0^2$$

is the velocity averaged over the cross section of the capillary. On the other hand, the mass is given by

$$m_e = 4\pi R_0^4 \rho_l l / 3 r_0^2 \tag{94}$$

Substituting (90) and (94) in (91), we obtain

$$f_r = \left(r_0/4\pi R_0\right) 3 p_{max}/z_{0\,max}\,\rho_l\,l \tag{95}$$

The damping force is

$$k\,\dot{z}_0 = \pi r_0^2\,\Delta p_c$$

where Δp_c is the pressure drop across the capillary. Since

$$\Delta p_c = 2\pi r_0 l \rho_l \nu_l \,|(\partial w/\partial r)_{r=r_0}|$$

and using (93), we have

$$k = 8\pi(R_0^4/r_0^4) l \rho_l \nu_l$$

$$\alpha = k/2\pi m_e f_r = (4 R_0\,\nu_l/r_0^3)\,(3\rho_l l z_{0\,max}/p_{max})^{1/2} \tag{96}$$

Figure 60 shows the results of calculations based on (95) and (96) for $R_0 = 1$ mm, $p_{max} = 10^3$ Pa, $z_{0\ max} = 0.025$ mm, $\rho_l = 1.2$ kg/m^3, and $\nu_l = 1.5 \times 10^{-5}$ m^2/s (air). For example, for $l = 1$ mm and $r_0 = 0.05$ mm, the resonance frequency turns out to be $f_r = 1.3$ kHz, whereas in the absence of the pulse transmission line, $f_r \simeq 10$ kHz for membrane thickness 5 μm and the same values of r_0 and p_{max} (see Fig. 20).

The damping coefficients of a system incorporating a pulse transmission line are much higher: for $r_0 = 0.05$ mm and $l = 4$ mm, the damping coefficient is $\alpha = 9$. However, when we consider these calculations *in toto*, we note that it is sometimes useful to mask part of the pressure transducer membrane in the case of measurements in gaseous media, since some of the deterioration in the dynamic properties can be compensated by a significant increase in spatial resolution [52].

It is clear from (95) and (96) that the damping introduced by the pulse transmission line depends on the density and viscosity of the liquid. For example, in water, f_r must be reduced by a factor of about 30 and α by a factor of about 3 (see Fig. 60).

CHAPTER
FIVE
ELECTRONICS FOR OPTICAL–FIBER TRANSDUCER

17. SOURCES OF RADIATION FOR OPTICAL–FIBER TRANSDUCERS

The radiation flux entering the return lightguides of optical-fiber displacement transducers determines their sensitivity and, consequently, has a direct influence on the resolution of velocity and pressure transducers.

In the case of pressure transducers, the source intensity can be quite moderate because both the feed and return lightguide bundles usually consist of many tens of fibers. This means that the measured light flux is relatively high even when glass membranes without reflecting coatings are employed. The factors that dictate the choice of the light source for a pressure transducer are mainly those of accessibility, cost, and scattering power.

Single-core lightguides, 0.01 – 0.02 mm in diameter, are used in velocity transducers, so that high-output sources are necessary to ensure good metrological characteristics. Unfortunately, high-intensity sources of coherent monochromatic radiation (lasers) are not suitable for this purpose because the displacement of the sensitive element is then of the order of the wavelength, and the output characteristics of the transducers have undesirable properties due to diffraction phenomena (see Section 3).

Sources of radiation used in optical-fiber velocity transducers must also satisfy certain further requirements, depending on application. First, the source must produce highly directional radiation because it is often used in combination with photodetectors and signal preamplifying circuits. Unless this is so, special optical coupling systems must be employed, which reduce reliability and degrade the working parameters of the measuring system. For the same reasons, it is inconvenient to have to cool the source by circulating a coolant such as water. On the other hand, photo-converters and other electronic devices are sensitive to temperature changes (see Section 18), so that one either must avoid appreciable heat releases or use cooling (thermostat) systems.

The spectral range of the source of radiation must match the transmission band of standard lightguides made from ordinary glass (0.5 – 1.3 μm). Quartz lightguides, which have a wider transmission band, are undesirable for technological reasons. The high-softening point of quartz glass means that one cannot use the simple fabrication technology described in Section 9 for optical-fiber displacement transducers. The source must also be matched to the spectral characteristic of the photodetector.

When static velocity and pressure distributions are examined, the time spent in one cycle of measurements can be quite long (up to 1000 s or more), so that instrumental parameters must, of course, be constant during this time. The characteristics of radiation sources (and of other components of the apparatus) must therefore be highly stable in time and have the minimum possible temperature dependence. It is desirable to use pulsed sources of radiation because low intensity light fluxes can then be measured with higher precision, and background effects can be eliminated. The latter factor is particularly important when the parameters of natural objects such as the atmosphere, the oceans, and so on, are examined.

Table 3

Source	Scatt. power, W	Radiated power, MW	Diameter, mm	Life, 10^6 hr	Temperature coefficient, %/K	Pulsed Operation
KGM6, 3 × 15 lamp	15	—	6	0.2	—	N/A
Photodiode AL-119	0.55	40	3	10	1–2.5	Up to 10^5 Hz
Photodiode AL-124	0.2	4–8	3	25	0.5–0.9	Up to 10^7 Hz
Photodiode AL-135	0.18–0.25	5–10	3	25	0.5–0.9	Up to 10^7 Hz

The above requirements suggest the use of two types of source, namely, the hot–filament lamp and the semiconductor source. The radiance of hot–filament lamps is proportional to the fourth power of the filament temperature, and is a maximum for the quartz halogen lamps manufactured in the Soviet Union (type KGM). The bulb temperature must be maintained at 500° C to ensure that the halogen reduction cycle operates satisfactorily for the evaporating tungsten wire. The minimum power dissipated by the lamp is then 15 W. A power dissipation level of 5 W is also possible for some such lamps, but the filament is then in the form of a thin wire rather than a helix. The filament vibrates and this modulates the light flux entering the lightguide. The KGM lamp requires water cooling and can therefore be used only when the maximum possible radiance without modulation is required.

Light–emitting semiconductor diodes have a somewhat lower (by 15 – 20%) light output than halogen lamps. However, they have many advantages such as small size, low power dissipation, long life, time and temperature stability, and possibility of pulsed operation. They are therefore recommended as light sources for miniaturized optical–fiber velocity transducers.

Table 3 lists some of the parameters of these light sources [44, 50].

The AL-124 and AL-135 sources are used in the velocity measuring equipment described below. The wavelength at maximum spectral intensity is then 0.86 – 0.88 μm, and this is well matched to the lightguide transmission band.

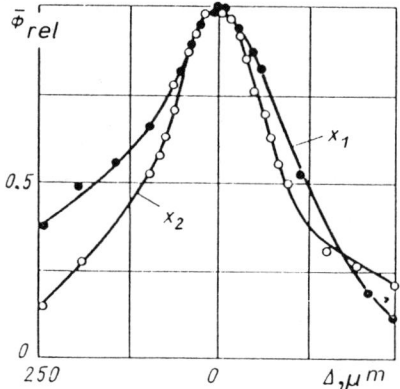

Fig.61 Light flux launched into a lightguide as a function of the distance from the center of the radiating area

The light-emitting diode has a small emitting area (0.3 × 0.3 mm) which means that it must be carefully adjusted relative to the center of the area because the light flux launched into the lightguide decreases rapidly with distance from this center. Figure 61 shows the measured relative light flux $\bar{\Phi}_{rel}$ launched into a lightguide 0.12 mm in diameter. The horizontal axis shows the distance from the center of the square radiating area along two mutually perpendicular directions (x_1 and x_2) parallel to the sides of the square.

18. PHOTODETECTING DEVICES

As noted in Section 17, the light flux arriving from the feed lightguide of the velocity transducer is very low (about 10^{-6} lm). The photodetector that transforms this flux into an electrical signal must therefore satisfy quite stringent requirements as regards sensitivity, noise level, temperature and time stability. Experience gained with velocity transducers working in conjunction with secondary-electron detectors suggests that the main source of error in velocity measurements is in fact the photodetector or, more precisely, the photodetector plus the preamplifier.

Photomultipliers are found to have the highest sensitivity among all the known photodetectors and therefore are widely used. However, the photomultiplier parameters are not very stable as functions of time. Even for the best examples, the phenomenon of photocathode ageing leads to a 3% change in sensitivity over six hours of continuous operation [24]. Signal drift is usually observed to be up to 10% over a few hours, which gives rise to appreciable errors in the case of prolonged measurements.

The electrons traverse relatively long paths in the photomultiplier, and this is the reason for the high sensitivity of the device to different types of magnetic field [59], which is another disadvantage of the photomultiplier. There is limited scope for the use of ferromagnetic shields for photomultipliers, since the weight of the photodetector system incorporating a photomultiplier and a ferromagnetic shield amounts to some tens of kilograms. Finally, the linear dimensions of the photomultipliers are relatively large, so that the velocity transducer cannot be physically combined with the photodetector. This means that long connecting lightguides have to be employed, which reduces the reliability of conversion and introduces significant practical limitations. Finally, the photomultiplier requires stabilized high-voltage supplies, which is not always convenient.

The operation and properties of semiconductor photodetectors using the internal photoelectric effect are described in detail in [45].

Studies of different photodetecting devices working together with

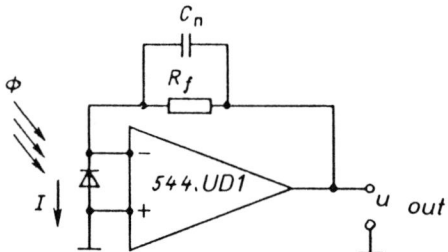

Fig.62 Block diagram of photodetector circuit

velocity transducers have shown that the most suitable device is the silicon photodiode with a current–voltage converter connected across it as shown in Fig. 62. For this arrangement,

$$u_{out} = IR_f + I_b R_f + u_b \qquad (97)$$

where I is the total photodiode current, R_f is the feedback resistor, I_b is the input current of the operational amplifier, and u_b is the bias voltage of the operational amplifier.

Temperature drift of the photodetecting device. The current–voltage characteristic of the photodiode operating as a photogalvanic device is described by [54]

$$I = \Phi S - I_s [\exp q u_d/kT - 1] \qquad (98)$$

where Φ is the light flux, S is the photodiode sensitivity,

$$I_s = B \exp(-E_s/kT) \qquad (99)$$

is the dark current [54], u_d is the voltage applied to the photodiode, q is the charge of electron, k is Boltzmann's constant, E_s is the band gap of the photodiode material, and B is a constant that depends on the photodiode fabrication technology, the degree of doping of the material and other factors.

It follows from (98) that the total photodiode current contains the useful signal ΦS as well as the error current

$$I_{err} = I_s \left[\exp qu_d/kT - 1\right] \qquad (100)$$

so that

$$u_{out} = \Phi S\, R_f - I_s \left[\exp qu_d/kT - 1\right] R_f + u_b + I_b\, R_f \qquad (101)$$

The last three terms in (101) represent the error in the measured light flux. This error could be taken into account in the form of an additive correction determined for $\Phi = 0$. However, the situation is complicated by the fact that the error depends significantly on temperature. Differentiating (97), and taking (98) and (100) into account, we obtain

$$\frac{du_{out}}{dT} = \frac{du_b}{dT} + \frac{di_b}{dT} R_f + \frac{dR_f}{dT} I_b - \frac{dI_{err}}{dT} R_f - I_{err} \frac{dR_f}{dT} \qquad (102)$$

The first three terms in (102) determine the components of the photodetector temperature drift introduced by the operational amplifier and the feedback resistor. Estimates based on information available about the operational amplifier and the feedback resistor suggest that the relative error is approximately 0.15%/K.

As far as the contribution of the photodiode to the temperature drift is concerned, it is represented by the last two terms on the right hand side of (102), and the penultimate term exceeds the last term by approximately an order of magnitude. Calculations and measurements show that for $T = 300$ K and a nominal $u_b = 20$ mV, we find that $dI_{err}/dT \sim 10$ pA/K and increases exponentially with increasing T and u_b. For the nominal value of the light flux intercepted by the velocity transducer, the photodiode current amounts to about 500 pA. Consequently, the relative experimental error due to the temperature dependence of the photodiode characteristics is 2%/K.

This error can be reduced by (1) reducing u_d by reducing u_b, (2) by selecting a photodiode with minimum I_s, (3) by parametrically compensating I_s for each photodiode by connecting it in opposition to a covered photodiode with a very similar I_s at the same temperature, and (4) by subtracting the error signal generated by a separate photodetector with a covered photodiode from the signal produced by the measuring photodetecting device. It is assumed that the values of T are the same for all the photodiodes.

A reduction in u_d is an effective way of reducing the error, and can be used together with the other methods enumerated above. It follows from the photodetector circuit that u_d can be determined with sufficient precision from the approximate formula

$$u_d = au_b + IR_f/A$$

where A is the gain of the operational amplifier and a is a constant.

It was confirmed experimentally that $u_b = 0$ and that u_d was proportional to the light flux. As Φ was increased from zero to the maximum value, it was found that u_d increased linearly, reaching 300 μV, i.e., roughly one half of u_d was due to u_b. Unfortunately, the photodetector temperature drift for $\Phi \neq 0$ cannot be fully compensated by subtracting the signal from an identical photodetector with a covered photodiode. However, the relative temperature drift error remains almost constant with increasing Φ because the useful signal increases. It follows that a reduction in u_d can be achieved by choosing an operational amplifier with low u_b and large A. However, this choice is restricted to commercially available operational amplifiers with output field-effect transistors. There is also little point in reducing u_b by balancing the operational amplifier because the temperature drift of the bias increases substantially (by 50 - 100 μV/K) for operational amplifiers with FET input [16].

The parametric compensation of I_s in each photodetector is inconvenient because this doubles the number of photodiodes in the system. For a two-component velocity transducer, with automatic adjustment of the radiation flux delivered by the lightguide, this method requires the use of six photodetectors. A more convenient procedure is to use one photodetector device with a covered photodiode and to subtract its signal from the signals due to the remaining three photodetector systems.

Noise characteristics of a photodetector. The error in the measured velocity is determined not only by the temperature instability of the photodetector, but also by its noise. Whereas the effect of noise is substantially reduced when the velocity and the Reynolds stresses are averaged over a particular interval of time, the measured root-mean-square velocity fluctuation is found to be higher when noise is present. The signal-to-noise ratio and the sensitivity of the above photodetector circuit increase in proportion to R_f. On the other hand, R_f is limited by the limited dynamic range of the operational amplifier ($\pm$ 15 V) and the presence of the parasitic capacitance $C_p \simeq 0.3$ pF of the feedback resistor. This is described by the following two upper limits:

$$R_f < 15/I_{max}$$

$$R_f < 1/(2\pi f_{lim} C_p)$$

where I_{max} is the maximum working current of the photodetector system and f_{lim} is the upper frequency limit of the velocity transducer.

The amplitude-frequency characteristic of the photodetector device for $R_f = 2.2$ GΩ is shown in Fig. 63.

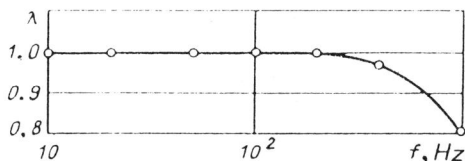

Fig.63 Amplitude–frequency charactistic of the photodetector for $R_f = 2.2$ GΩ

148   ELECTRONICS FOR TRANSDUCERS

Suppression of external electromagnetic interference. In the equipment described above, the photodetector unit is connected to the remainder of the system by a 2-m cable. External interference induced in this cable is suppressed by the differential amplifier illustrated in Fig. 64. The figure also shows the circuit used to compensate the temperature drift of the photodetector system. The compensated voltage u_T from the photodetector output with covered photodiode is amplified by the differential amplifier and is subtracted from the voltages u_1 and u_2 in the measuring channels and the voltage u_f in the feedback channel of the automatic radiation-flux control system.

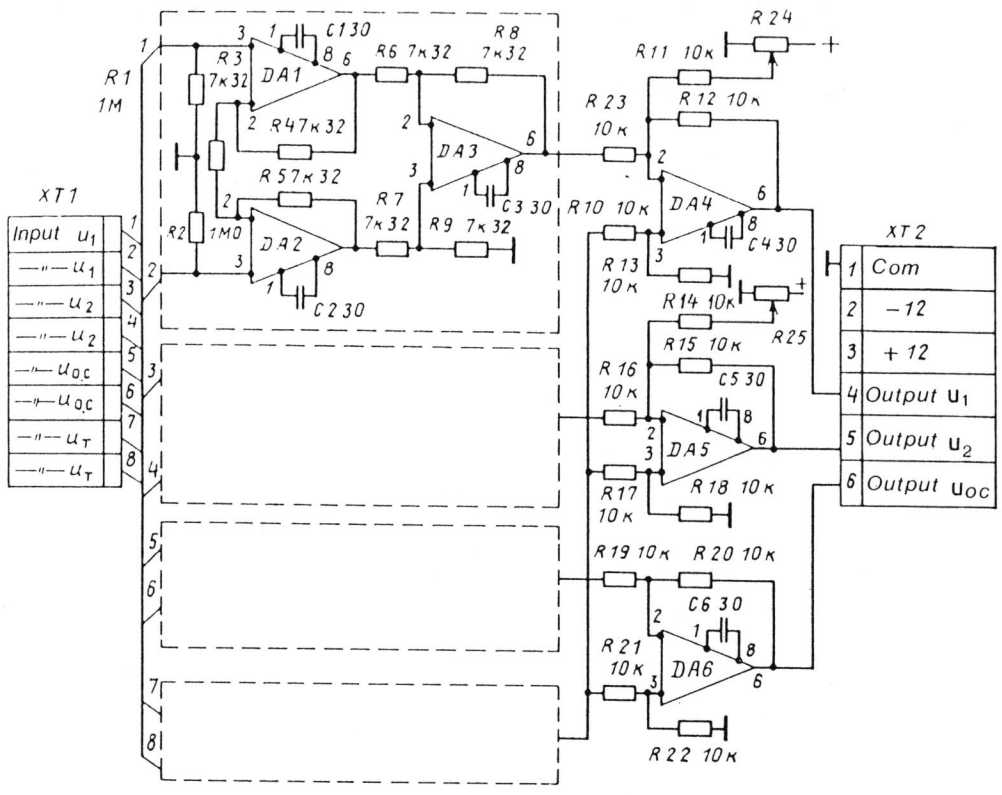

Fig.64 Block diagram of the differential amplifier

The potentiometers R24 and R25 are used to compensate the signal corresponding to zero fluid flow rate.

Protection of the equipment from industrial interference that penetrates via the mains supply sources is also a relatively difficult problem. Compensated voltage stabilizers cannot handle short pulses of mains interference because of the limited intrinsic passband. Mains filters produced commercially are not adequate either. The source of supply for units that are particularly sensitive to mains interference is therefore usually based on a voltage converter.

The greatest hazard is presented by interference that enters in measuring circuits via the parasitic capacitance of the supply transformer, and steps must therefore be taken to ensure that the transformer has a large coil and a very small capacitance between the primary and secondary. Multibranch output filters are used to reduce pulsed converter interference down to the level of a few millivolts.

Tests made on the source of supply using an interference simulator showed that mains interference pulses of length between 10 and 100 μs can be reduced by 60 dB.

Stabilization of radiation flux. The radiation power generated by the source decreases with increasing crystal temperature. Moreover, there is a linear fall in the radiation power with time, which is due to degradation of the source working medium. Any change in the radiation flux introduced into the feed lightguides of the velocity transducer produces a relative error in the measured velocity that is equal to the relative change in the radiation flux.

This error is reduced by an automatic radiation flux controller that maintains the flux at a constant level. It diverts part of the flux to a feedback lightguide glued to the illuminating lightguides of the velocity transducer. It is used in conjunction with photodetectors that are identical with those employed in the measuring channels.

The automatic light flux control system operates effectively only when the photodetector characteristics are temperature-stabilized as described above. Unless this is done, a spurious signal will be produced by the temperature drift.

19. THE MICROPROCESSOR

As an example, consider the microprocessor used in measurements of the two components of instantaneous velocity in a flow of liquid metal (mercury). This system must implement the algorithm for calculating the velocity components from the two signals produced by the optical-fiber displacement transducers described in Section 10.

Measurements of liquid metal flow velocity give rise to certain difficulties when standard equipment normally used for measurements on air flows is employed. One of the distinctive features of this situation is the presence of ultra-low-frequency velocity fluctuations. Measurements of the average hydrodynamic quantities under these conditions require long integration times, and this obviously gives rise to instrumentation difficulties.

Measurement of averages. Two methods are used to measure the average signal over long intervals of time: one is the analog method and the other the digital method.

An analysis [16] of an analog integrator incorporating good quality operational amplifiers with FET input suggests that the integrator error amounts to about 10% over an integration time of about 1000 s. The digital method is more precise (up to 0.01%), and the precision is determined only by the error in the analog-to-digital convertor, and is independent of the integration time. However, the measuring system can only be used in conjunction with a data acquisition system connected to a computer, which restricts its operational possibilities. Moreover, digital systems of this type are much more expensive than analog systems.

A composite analog/digital method of evaluating averages was employed to process the output signal from optical-fiber velocity and pressure transducers. This method is comparable in complexity with the analog method but, in contrast to the latter, it introduces an error of the order of 0.1%, which is independent of the integration time. The digital integrator used in this equipment is based on the voltage-to-frequency converter (VFC) whose block diagram is illustrated in Fig. 65. It is shown in [48] that the contents of the register of counter 4 which is connected directly to the VFC output 1, is equal to $\int_0^T u(\tau)d\tau$ after the time T. The regulated frequency divider (RFD) 2 whose division factor is equal to T (in seconds) is connected between the VFC and the counter so that the counter register records the number $<u> = (1/T)\int_0^T u(\tau)d\tau$, i.e., the average output signal $u(\tau)$ over the time interval T. When the integration time is specified arbitarily (for example, by some external effect), then the controlling frequency divider can be replaced by a clock, and the contents of the register are divided by T.

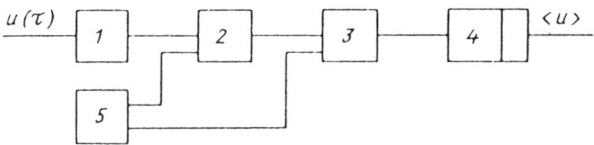

Fig.65 Block diagram of integrator: 1 – VFC, 2 – RFD, 3 – timer or external gating device, 4 – counter with register, 5 – gating device

This system can be used to integrate an input signal over an interval T with an accuracy limited only by the accuracy of the VFC, which can be judged by the certificate of performance issued by the manufacturers. In the particular case of the mass-produced converter 1108 PP1 [18], the certificate shows the following:

Conversion nonlinearity, % 0.01

Temperature drift of the conversion coefficient, %/K	0.015
Response time to a 10 V voltage step, μs	40
Bias voltage for the input operational amplifier, mV	4

To minimize the influence of ground interference, the digital part of the integrator, which is directly connected to external instrumentation, is decoupled galvanically from the measuring part by an optoelectronic inverter.

The integration time can be as long as desired because the error is independent of this time. This particular instrument has nine values of integration time ranges (between 3 and 1000 s). In addition, the integrator can be started and stopped by an external signal. The integration time is then determined by the time interval between start and stop pulses.

Measurement of the two velocity components and of their moments. As shown in Section 10, the velocity transducer produces output signals u_1 and u_2 that are related to the velocity components w_1 and w_2 by (23) and (24) from which it follows that

$$w_1 = A u_1^m$$

$$w_2 = B w_1 u_2 / u_1 \tag{103}$$

In a turbulent flow, w_1 and w_2 are random functions of time, so that it is sensible to measure only the time-averaged flow parameters. If w_1 is the velocity component parallel to the flow direction and w_2 is the component perpendicular to it, we have

$$w_1(\tau) = <w_1> + w'_1(\tau)$$

$$w_2(\tau) = w'_2(\tau)$$

$$<w_2> = 0 \tag{104}$$

where $<w_1>$ is the time average of w_1 and w'_1, w'_2 are the turbulent fluctuations in the velocity components.

The equipment described above can measure the following quantities:

$$<w_1> = (1/T) \int_0^T w_1(\tau) \, d\tau \tag{105}$$

$$\text{RMS } w'_1 = \left[(1/T) \int_0^T [w'_1(\tau)]^2 \, d\tau \right]^{1/2} \tag{106}$$

$$\text{RMS } w'_2 = \left[(1/T) \int_0^T [w'_2(\tau)]^2 \, d\tau \right]^{1/2} \tag{107}$$

$$<w'_1 w'_2> = (1/T) \int_0^T [w'_1(\tau) w'_2(\tau)] \, d\tau \tag{108}$$

The last of these quantities is called the Reynolds friction stress and is typical of turbulent momentum transfer.

Measurement of $<w_1>$, $<w'_1 w'_2>$, and the root-mean-square value of w'_2 is relatively simple because it follows from (104) that

$$\text{RMS } w'_2 = \text{RMS } w_2$$

$$<w'_1 w'_2> = <w_1 w_2> \tag{109}$$

and the measurement process reduces to simple operations defined by (103), (105), (107), (108), and (109). When the root-mean-square value of w'_1 is determined, it must be remembered that

$$w'_1(\tau) = w_1(\tau) - <w_1> \quad \text{RMS } w'_1 = \left\{ (1/T) \int_0^T [w_1(\tau) - <w_1>]^2 \, d\tau \right\}^{1/2}$$

The ultra-low-frequency spectrum of velocity fluctuations that is typical for liquid-metal flows under laboratory conditions complicates the instrumental implementation of calculations based on the last equation because it is difficult to isolate the constant component $<w_1>$ directly during the measurement of RMS w'_1. In particular, there is no point in using filters because this would also remove the energy-carrying low-frequency components of the spectrum.

We have therefore based our measurements of RMS w'_1 on the following obvious relations

$$(\text{RMS } w'_1)^2 = <w_1^2> - <w_1>^2$$

By integrating $w_1(\tau)$ and $w_1^2(\tau)$ over the interval T we obtain information on both $<w_1>$ and RMS w'_1, so that RMS w'_1 can be calculated in digital form after the integrator:

$$\text{RMS } w'_1 = \left\{ (1/T) \int_0^T w_1^2(\tau)\, d\tau - \left[(1/T) \int_0^T w_1(\tau)\, d\tau \right]^2 \right\}^{1/2} \quad (110)$$

Figure 66 shows a scheme for evaluating $<w_1>$, RMS w'_1, RMS w'_2 and $<w'_1 w'_2>$ using the above algorithm. In this scheme, I1–I4 are integrators, AD1, AD2 are adders, and A2–A6 are analog multipliers operating on the four input voltages u_x, u_y, u_z, and u_R in accordance with the following expression:

$$u_{out} = \left[(u_x + u_1)(u_y + u_2) \right] / \left[(u_z + u_3) - (u_R + u_4) \right] \quad (111)$$

The multipliers are based on the well-known logarithm – antilogarithm scheme [55]. The following parameters can be attained when these multipliers are suitably adjusted:

Relative error in the range 1 mV – 5 V, %	≤ 0.2
Nonlinearity of characteristics, %	0.1
Temperature coefficient of output voltage (20–60 °C), %/K	0.03

When such high precision is not essential, it is possible to use the simple multipliers (type 525 PS2) [1] with a nonlinearity of 0.5 – 2% and an error of about 1%.

The absolute value of the signal is generated by ABS1 and ABS2 (analog multipliers of the logarithm – antilogarithm type work only in one quadrant).

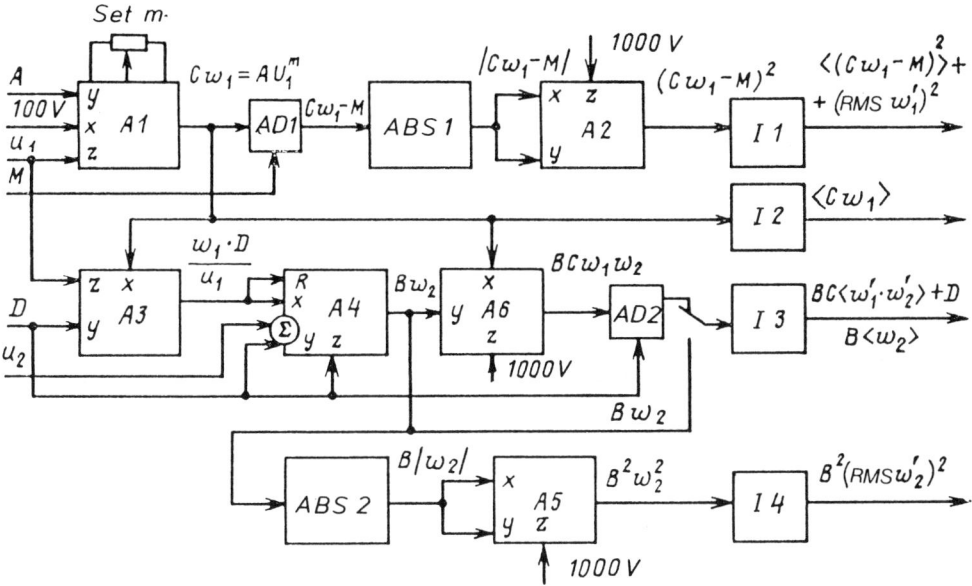

Fig.66 Block diagram of a computing device

The analog linearizer A1 performs the following operation on the three input voltages u_x, u_y, and u_z:

$$u_{out} = u_y(u_x/u_z)^m \tag{112}$$

where the exponent m may vary with the value of n in (23). A ten-turn potentiometer is used to set the value of m. The dependence of m on the angle α of the rotation of the potentiometer is linear, so that two or three points are sufficient to determine the actual function $m(\alpha)$. As noted in Section 10, the value of n for velocity transducers used in flowing mercury lies in the range 1.5–1.8. It is therefore convenient to confine the variation in the exponent $m = 1/n$ to the range $0.500 - 0.666$. Under these conditions, the value of m can be set to within 0.001. With careful calibration, one can easily achieve a nonlinearity of the transfer characteristic of not more than 0.05% and an error of about 0.1%. The temperature coefficient of the output voltage does not exceed $0.03\%/K$.

The structure of the analog linearizer is similar to that of the above analog multiplier [55].

The symbol Σ indicates additions at the multiplier input. Since w_2 has a variable sign, the signal is shifted by the amount D, equal to one-half of the full scale, at the input of the multiplier A4. A voltage of 1000 V is applied to the unused multiplier inputs. The constants A, D, M are specified in the form of voltage is generated by sources with voltage temperature coefficients of not more than $0.01\%/K$.

The coefficients B, C, and m are determined during the calibration of the transducer.

The apparatus is also capable of evaluating $<w_2>$ for which w_2 is introduced in analog form, and, if necessary, is connected to the integrator I3. This is done to ensure accurate orientation of the velocity transducer relative to the flow (see Section 10). The condition $<w_2> = 0$ is achieved by rotating the velocity transducer around the axis of its sensitive element.

It is clear from (110) that $(\text{RMS } w'_1)^2$ is a relatively small difference between two large quantities, i.e., it is subject to a large relative uncertainty. Evaluating this uncertainty in the usual way, we obtain

$$\delta \text{RMS } w'_1 \sim \frac{\Delta}{2}\left[1 - \left(\frac{<w_1>}{\text{RMS } w'_1}\right)^2\right]$$

where Δ is the relative uncertainty associated with the operations of squaring and averaging. It is clear that, when RMS w'_1 is small, the uncertainty is large. To reduce this uncertainty, the squaring operation is preceded by shifting the signal w_1 by the amount $M \simeq <w_1>$. The relative uncertainty then becomes

$$\delta \text{RMS } w'_1 \sim \frac{\Delta}{2} \left[1 - \left(\frac{<w_1> - M}{\text{RMS } w'_1} \right)^2 \right]$$

i.e. it is close to $\Delta/2$.

Pressure measurement. The apparatus described above can be used with minor modifications to measure the pressures p_1 and p_2 at two points in a turbulent flow. In view of the nature of the calibration curve of the pressure transducer (see Fig. 22), it is best to employ a linearizer based on a piece-wise linear or polynomial approximation [55]. Measurements of the average pressure $<p_i>$, the root-mean-square pressure fluctuations RMS p'_i, and the correlation $<p'_1, p'_2>$ are not signficantly different from the measurements of $<w_1>$, RMS w'_1, and $<w'_1, w'_2>$.

CONCLUSION

The future prospects for optical-fiber pressure and velocity transducers are determined by properties such as high sensitivity, good dynamic characteristics, ease of fabrication, and ability to work in conjunction with optical-fiber data telemetering channels.

As far as the optical-fiber velocity transducers are concerned, there are areas in which these transducers cannot compete with traditional methods. Measurements on electromotive flows have already been mentioned in Section 10. The second application of this kind is to free turbulent convection near a hot surface. Here, the optical-fiber velocity transducer can be used to measure not only the mean velocity field, but also the statistical parameters of the field. It is also important to remember that velocity transducers can readily be transformed into transducers for the tangential friction stress on the surface of a body placed in the flow. This can be done with the elastic sensitive element of the velocity transducer, augmented with a receiving area element placed next to the surface. The optical-fiber/optical-displacement transducer is then found to require practically no modification.

Optical-fiber pressure transducers are beginning to be used in different technologies. For example, they have been used to measure, for the first time, the rapidly varying pressure in the lubricant of high-speed bearings [63]. Optical-fiber transducers have promising applications in low flow-rate measurement, vacuum gauges, in studies of recombination processes in which a reduction in particle concentration leads to a rapid pressure drop in the system, and so on. Because of their relative simplicity, low cost, and small size, these transducers may well find routine applications in automatic control systems.

REFERENCES

1. S.V. Yakubovskii (ed.), *Handbook of Analogue and Digital Integrated Microcircuits* [in Russian], Radio i svyaz', Moscow, 1984

2. V.G. Zhilin and V.V. Osipov, "Optico-mechanical velocity transducer for a transparent medium" Soviet Patent No. 463908; *Otkrytiya. Izobreteniya*, No.10 (1975)

3. V.G. Zhilin, V.P. Ogorodnikov, and V.V. Osipov, "One-component optico-mechanical velocity transducer" Soviet Patent No.617720; *Otkrytiya. Izobreteniya*, No. 28 (1978)

4. V.G. Zhilin, V.P. Ogorodnikov, and V.V. Osipov, "Two-component optico-mechanical velocity transducer" Soviet Patent No.684448; *Otkrytiya. Izobreteniya*, No. 33 (1979)

5. V.A. Atsyukovskii, *Capacitive Displacement Transducers*, Energiya, Moscow, Leningrad, 1966

6. S.T. Pai, *Fluid Dynamics of Jets*, Van Nostrand, 1954

7. M.Born and E. Wolf, *Fundamentals of Optics*, Pergamon Press, Oxford, 1980

8. G.G. Branover, N.M. Slyusarev, and E.V. Shcherbinin, "Some results of measurements of turbulent velocity fluctuations in the flow of mercury in the presence of a transverse magnetic field", *Magnitnaya gidrodinamika*, No. 1, p.33 (1965)

9. B.M. Budak, A.A. Samarskii, and A.N. Tikhonov, *Collection of Problems on Mathematical Physics*, Pergamon Press, 1964

10. G.K. Batchelor, *Introduction to Fluid Dynamics*, Cambridge University Press, 1967

11. G.A. Gachechiladze, "An instrument for measuring the fluctuation

velocity components in plane turbulent flow", in: *New Methods of Measurements and Instrumentation for Hydraulic research* [in Russian], V.V. Zvonkov (editor), Academy of Sciences of the USSR, p.85, 1961

12. E.A. Genina, "Electrical devices for hydraulic research", *ibid.*, p.117

13. V.G. Zhilin, Yu. P. Ivochkin, V.P. Ogorodnikov, and V.V. Osipov, "A two-component optico-mechanical velocity transducer for studies of turbulent liquid-metal flows", *Teplofizika vysokikh temperatur*, Vol. 20, 1164 (1982)

14. B.S. Petukhov, V.G. Zhilin, Yu. P. Ivochkin, et al., in: *Convective Heat Transfer − New Methods and Instrumentation for Hydraulic Research* [in Russian], B.S. Petukhov (editor), Academy of Sciences of the USSR, p.181 (1982)

15. M.M. Didkovskii, B.M. Egidis, and N.G. Poznyaya, "A sensor for the simultaneous measurement of two instantaneous velocity components", *ibid.*, p.75

16. J.Dostal, *Operational Amplifiers*, Elsevier, 1981

17. Yu.N. Dubintsev and B.S. Rinkevichus, *Methods of Doppler Laser Anemometry* [in Russian], Nauka, Moscow, 1982

18. V.B. Dychakovskii, S.I. Kobzar', and S.L. Sud'in, "The KR1108PP1 voltage-frequency-voltage transducer", *Elektronnaya Promyshlennost'*, USSR Ministry of Electronics, No. 6, p.28 (1984)

19. V.G. Zhilin, V.P. Ogorodnikov, and V.V. Osipov, "Dynamic characteristics of optico-mechanical velocity sensors for transparent media", *Teplofizika vysokikh temperatur*, **18**, 387 (1980)

20. V.G. Zhilin, "On the possibility of using optical-fiber velocity transducers for measurements near the wall of a hot pipe", *ibid.*, **22**, p.769 (1984)

21. V.G. Zhilin, Yu.P. Ivochkin, and A.A. Oksman, "On the absence of

the influence of a transverse magnetic field on the readings of an optical-fiber velocity transducer in a liquid-metal", *Teplofizika vysokikh temperatur,* **22**, 1024 (1984)

22. V.G. Zhilin, Yu.P. Ivochkin, and A.A. Oksman, "A method of measuring the two components of velocity in a turbulent flow of a liquid metal using an optical fiber velocity transducer", *ibid.,* **22**, 1178 (1984)

23. A.F. Zak and Yu.P. Man'ko, "The influence of temperature on the deformation and strength of glass fibers", *Zhurnal tekhnicheskoi fiziki,* **24**, 1983 (1954)

24. B.V. Kantsel'son, A.M. Kalugin, and A.S. Larionov, *Vacuum Tube Electronic and Ionic Devices* [in Russian], Energiya, Moscow, 1970

25. L.G. Kit, "Measurement of turbulent velocity fluctuations using the conductive anemometer with a three-electrode transducer", *Magnitnaya gidrodinamika,* No. 4, 41 (1970)

26. G. Comte-Bellot, "Turbulent flow between two parallel walls" Pub. Sci. et Tech. du Min. de l'Air (France) No.419, 1965, translated as Aero Res. Council Paper No. 31609, 1969

27. B.P. Korop, "Measurement of the velocity of fluids by electric methods", in: *New Methods of Measurements and Instrumentation for Hydraulic Research* [in Russian], V.V. Zvonkov (editor), Academy of Sciences of the USSR, p.65, 1961

28. L.M. Korsunskii, "The region of spatial averaging by a conductive anemometer with an external uniform magnetic field", *Magnitnaya gidrodinamika,* No. 4, 148 (1974)

29. M.P. Lisitsi, L.I. Berenzhinskii, and M.Ya. Valakh, *Fiber Optics* [in Russian], Tekhnika, Kiev, 1968

30. W.H. MacAdams, *Heat Transmission,* McGraw Hill, 1954

31. D.J. Malcolm, "Hot-wire anemometer measurements in

nonstationary magnetohydraulic flows of liquid metals using insulated platinum probes", *Magnitnaya gidrodinamika*, No.2, 55 (1970)

32. L.G. Markova, "Frequency characteristics of glass insulated fiber-sensor hot-wire anemometer for measurements in liquid metals", *Magnitnaya gidrodinamika*, No. 3, 141 (1974)

33. M.A. Mikhalev, "An instrument for measuring velocity fluctuations using a wire transducer", in: *New Methods of Measurements and Instrumentation for Hydraulic research* [in Russian], V.V. Zvonkov (editor), Academy of Sciences of the USSR, p.80, (1961)

34. B.Z. Mikhlin, *High-frequency Capacitive and Inductive Transducers* [in Russian], Gosenergoizdat, Moscow-Leningrad, 1960

35. Kh.E. Kalis, A.B. Tsinober, A.G. Shtern, and E.V. Shcherbinin, "Flow past a circular cylinder in an electrically conducting fluid in a transverse magnetic field", *Magnitnaya gidrodinamika*, No. 1, 18 (1960)

36. V.P. Ogorodnikov and V.V. Osipov, "Influence of a magnetic field on the calibration of an optical-fiber velocity trasducer", *Teplofizika vysokikh temperatur*, **22**, 808 (1984)

37. V.G. Zhilin, V.P. Ogoroenikov, V.V. Osipov, and B.S. Petukhov, *ibid.*, **14**, 834 (1976)

38. V.G. Zhilin, Yu.P. Ivochkin, V.P. Ogorodnikov, and V.V. Osipov, "Optico-mechanical pressure transducers", *ibid.*, **17**, 1064 (1979)

39. L.L. Pal and L.A. Tepaks, "Electro-optical method of measuring the dynamic characteristics of a flow using a photoresistor", in: *New Methods of Measurements and Instrumentation for Hydraulic research* [in Russian], V.V. Zvonkov (editor), Academy of Sciences of the USSR, p. 133, (1961)

40. R.H. Pahler and A.S. Roberts Jr. "Development of an optical pressure transducer with a fiber lightguide" [translated from English], Trans. ASME, 1977

41. G.S. Pisarenko, A.P. Yakovlev, and V.V. Matveev, *Handbook on the Resistivity of Materials* [in Russian], Naukova dumka, Kiev, 1975

42. I.A. Platnieks, "Comparison of the hot-wire anemometer with conductive methods of measuring the velocity parameters of a mercury flow in a transverse magnetic field", *Magnitnaya gidrodinamika*, No.3, 140 (1971)

43. I.L. Povkh, *Aerodynamic Experiment in Machine Construction* [in Russian], Mashinostroenie, 1974

44. A.V. Bayukov et al., in *A Handbook of Semiconductor Devices: Diodes, Thyristors, and Optoelectronic Devices* [in Russian], N.N. Goryunov (editor), Energoizdat, 1982

45. I.D. Anisomova et al., in *Semiconductor Photodetectors: Ultraviolet, Visible, and Near-infrared Ranges* [in Russian], V.I. Stafeev (editor), Radio i svyaz', 1984

46. A.F. Polyakov, and S.A. Shindin, "Hot-wire anemometer measurements of average velocity in the immediate proximity of a wall", *Inzhenerno-fizicheskii zhurnal*, **36**, 985 (1979)

47. G.L. Popova, "High-sensitivity small-size, and miniaturized pulsed-pressure probes", Tr. Tsentral'nogo in-ta aviats. motorostroeniya, No. 563 (1973)

48. G.M. Petrov et al., in: *Data Processing in Alphanumeric Computing Devices and Systems* [in Russian], G.M. Petrov (editor), Mashinostroenie, 1973

49. E.K. Rabkova and L.K. Tubashov, "Automatic hydraulic system and measurement of flow parameters using electronic sensors", *Novye metody izmerenii i pribory dlya gidravlicheskikh issledovanii* [in Russian], Izd-vo AN SSSR, 1961, p.112

50. N.A. Gorbatov, A.A. Vilisov, V.M. Sheludkov and K.B.

Okhmanovich, "Light diode for internal fiber optic probes", in: *Tekhnika sredstv svyazi*, Ser. Vnutriob'ektovaya Zvyaz', No. 1, 154–160 (1984)

51. T.E. Siddon and H.S. Ribner, "Foil sensor for measuring the transverse component of turbulent fluctuations" [translation from English], *Raketnaya teknika i kosmonavtika*, No. 4, 217 (1965)

52. A.V. Smolyakov and V.M. Tkachenko, *The Measurement of Turbulent Fluctuations*, Springer, 1983

53. R.M. So, "The influence of shear in a flow during measurements with a hot-wire anemometer" [translation from English], Trans.ASME, Series D, No. 4, 288 (1974)

54. N.A. Soboleva and A.E. Melamid, *Photoelectric Devices* [in Russian], Vysshaya shkola, Moscow, 1974

55. D.H. Scheingold (ed.), *Nonlinear Circuits Handbook*, Analog Devices Inc., 1976

56. A.M. Trokhan, "Optical fiber pressure sensor", *Izmeritel'naya tekhnika*, No. 6, 42 (1988)

57. A.P. Filippov, *Oscillations of Deformable Systems*, Mashinostroeniye, Moscow, 1970

58. A.B. Tsinober, *Magnetohydrodynamic Flow Past Bodies* [in Russian], Zinatne, Riga, 1970

59. N.O. Chechik, S.M. Fainshtein, and T.M. Lifshits, *Photomultipliers* [in Russian], Gostekhizdat, 1954

60. H. Schlichting, *Boundary Layer Theory*, McGraw–Hill, 1968

61. P.V. Novitskii (ed.), *Electrical Measurements of Nonelectrical Quantities*, [in Russian], Energiya, Leningrad, 1975

62. V.V. Boyarevich, Ya.Zh. Freiberg, E.I. Shilova and E.V. Shcherbinin,

Electromotive Flows [in Russian], Zinatne, Riga, 1985

63. I.S. Yavelov, "Measurement of pressure in bearings", *Mashinovedenie*, No. 4, p.104 (1981)

64. M.L. Polyani, "Flexible Optical Probe", US Patent No. 3068739

65. C.R. Hazell and S.L. Engel, "A fiber optical angular displacement transducer", *J. Sci. Instr.* 2, 110 (1969)

66. T.R. Hsu, R.G. Moyer, and F.B. Banks, "A high temperature fiber optical displacement probe", *ibid.*, No. 12, 1132 (1962)

67. W.Krauss, *Messungen des Temperatur und Geschwindigkeit feldes bei freier Konvektion*, Karlsuhe, Braun, 1955

68. G.W. Margerum, *Fiber Optic Pressure Transducer*, M.S. Thesis, Naval Postgraduate School, Monterey, California, 1972.

INDEX

Anemometer:
   conductive, 65–67
   hot-wire, 2, 5, 23, 41, 61, 65
   laser Doppler, 61

Biot number, 94
Blood pressure measurement, 9

Cantilever:
   conical, 72
   oscillations, 111

Damping of oscillations, 103, 116
Diffraction by an edge, 21
Drag coefficient, 1, 42–43, 56

Flow:
   electromotive, 64
   liquid-metal, 75
   Stokes past cylinder, 41
   turbulent, 47, 53
   with velocity and temperature gradients, 90–94

Influence function, 63, 71, 74

Karman vortex, 42
Krylov function, 110

Lightguide, 9, 11–15, 23, 50

Magnetic field, 77, 78
Manometer:
   liquid filled, 7
Mass, associated, 132
Membrane:
   deflection, 34
   glass, 33, 35–36
   reflection coefficient, 33
   resonance frequency, 130
   thickness measurement, 37

Newton's rings, 37

Oscillations:
   with damping, 102, 116
   of sensitive element, 97–101

Photodetector:
   adjustment, 69
   circuit, 144
   noise, 147
   output analysis, 18, 26
   solid angle, 19
   temperature drift, 144–147
Photomultiplier, 6, 54
Piezoceramics, 8
Pitot tube, 5, 38
Prandtl, 65, 93

Reflection coefficient, 33
Refractive index, 11
Resolution:
   spatial, 60, 86–88
Resonance frequency, 12, 130–132
Reynolds number, 1–2, 41–43, 51

Sensor:
   velocity and pressure, 11

Transducer:
   calibration, 55–59
   capacitive, 8
   displacement, 7, 11, 15, 17, 49, 83
   drag on head, 2
   dynamic characteristics, 97, 107, 122, 130
   electrolyte, 3
   error, 128–130
   head, 1, 5
   magnetic field effect, 78–79
   membrane, 9, 29, 34, 39
   miniaturization, 5
   for opaque fluid, 68, 122–123
   optical-fiber, 6–9, 15, 29
   piezoelectric, 8
   pointer, 70

Transducer (*Cont.*):
    pressure, 7–9, 29, 33, 38
    radiation source, 139
    resonance frequency, 123
    sensitive element, 3, 41, 44, 46, 71, 97, 119
    sensitivity, 31–33, 62, 71
    for transparent fluids, 49, 107–110
    types, 1–3
    velocity, 4, 41, 54, 69

Velocity:
    fluctuations, 4, 42
    fluid, 4
    transducers, 41, 47, 69, 77, 112

Viscosity, 1

Young's modulus, 32, 47, 98